THE THEORY
THAT CHANGED
EVERYTHING

THE THEORY
THAT CHANGED
EVERYTHING

"ON THE ORIGIN OF SPECIES"

AS A WORK IN PROGRESS

PHILIP LIEBERMAN

COLUMBIA UNIVERSITY PRESS

New York

Columbia University Press
Publishers Since 1893
New York   Chichester, West Sussex
cup.columbia.edu
Copyright © 2018 Columbia University Press

Library of Congress Cataloging-in-Publication Data
Names: Lieberman, Philip, author.
Title: The theory that changed everything : "On the origin of species"
as a work in progress / Philip Lieberman.
Description: New York : Columbia University Press, 2017. |
Includes bibliographical references and index.
Identifiers: LCCN 2017007542 (print) | LCCN 2017040837 (ebook) |
ISBN 9780231545914 (e-book) | ISBN 9780231178082 (cloth : alk. paper)
Subjects: LCSH: Darwin, Charles, 1809–1882. On the origin of species. |
Darwin, Charles, 1809–1882—Influence. | Evolution (Biology) | Natural selection.
Classification: LCC QH365.O8 (ebook) | LCC QH365.O8 L54 2017 (print) |
DDC 576.8/2092 [B] —dc23
LC record available at https://lccn.loc.gov/2017007542

Columbia University Press books are printed on permanent and
durable acid-free paper.
Printed in the United States of America

Cover design: Noah Arlow

To Marcia, the love of my life

Little lamb who made thee.

Dost thou know who made thee?

—William Blake, 1789

CONTENTS

PREFACE

The more we learn, the more it becomes evident that the insights and observations conveyed in *On the Origin of Species* continue to guide our understanding of biology, evolution, and how we might act in matters of general interest, some bearing on our very existence.

Charles Darwin was a keen, curious observer who possessed a quality of mind that enabled him to see the connections among seemingly unrelated natural phenomena and human endeavors. His voyage on HMS *Beagle* was, as he put it, "the most important event in my life, and has determined my whole career." Darwin came in contact with the indigenous people of Tierra del Fuego, who lived in circumstances far removed from those of his prosperous family and acquaintances, but he nonetheless discerned the qualities that made it clear to him that all people are brothers and sisters. He saw animals and plants outside the range of what is found in England and Scotland and the fossil remains of creatures that no longer live on earth.

When Darwin first boarded the *Beagle* on December 27, 1831, he firmly held to the belief that God had created the world and all its forms of life. On his return to England four years later, he started on the journey in his mind's eye that led to the theory that changed how we view the evolution of all forms of life and their place in nature. His theory was controversial from the day of its publication in 1859 and remains so today. Darwin departed from the view that God had created the world and all forms of life in a week or else had set in motion a master plan that over time produced new forms of life, extinguishing others. The course of evolution instead was happenstance based on variation and chance. No two beings are identical. Variation always marks the creatures or plants that constitute a species, including the individuals who constitute the human species, *Homo sapiens*. Darwin did not know how or why variations occurred—why some roses were redder than others or why some people were taller or could run faster—but he thought that natural selection was the primary mechanism for the "transmutation" of species: evolution.

Natural selection is a simple concept. Any heritable variation that results in a living organism's having more descendants tends to be preserved, providing a "selective advantage" in the "struggle for existence." However, ruthless competition and brute strength were not the keys to success in the Darwinian struggle for existence, a concept that has been misinterpreted and misused to justify exploitation and greed. Cooperation yields benefits, and, as Darwin stressed, small changes can drive natural selection. One of the incremental results of natural selection is evident on the shelves of your supermarket. If your ancestors lived in a culture that herded cows, sheep, or goats, you will most likely skip the lactose-free selections because natural selection has conferred on you adult lactose tolerance. Being able to add milk to our diet conferred an advantage in the "struggle for existence"—making some individuals and their children more likely to survive. If your ancestors did not live in such a dairy culture, you may need to purchase lactose-free products.

However, Darwin was wrong in thinking that the pace of natural selection is always slow and gradual. Had he remained in the Galapagos for a decade rather than a few weeks, he might have observed changes occurring in the finches' beaks as they adapted to harder- or softer-shelled nuts. But Darwin pointed out that natural selection wasn't the only process that drove evolution. "Recycling," adapting an organ to do something new, could abruptly change the course of evolution. And he borrowed from the grandfather he had never met, Erasmus Darwin. Erasmus, in his book *Zoonomia*, published decades before Darwin was born, studied the development of animals from their embryological state onward, which to both Erasmus and Charles Darwin demonstrated the continuity of evolution.

Erasmus also thought that "acquired features," direct effects induced by the environment, could be inherited—the "just-so" story of how giraffes got long necks because they stretched to reach leaves on higher tree branches. Charles Darwin also borrowed that concept. In the twenty-first century, we're seeing that it isn't always a just-so story. I will attempt to demystify current "Evo-Devo" studies aimed at identifying transcriptional and epigenetic factors coded in your DNA that determine why you don't look or think like a chimpanzee, even though you share virtually all of its genes. Epigenetic effects, for example, caused the grandchildren of women from nineteenth-century Sweden whose diet suddenly changed from famine to feast, or the reverse, to die thirty years sooner than other people.

I have organized my book as though it were the script of a documentary film aimed at presenting Charles Darwin's insights. The "talking head," whenever possible, is Darwin, in passages drawn from *On the Origin of Species*, the short autobiography that he wrote for his children and grandchildren, and his letters. The exotic settings that Darwin encountered throughout South America and in the Galapagos Islands set his mind in motion, but he saw parallels and conducted experiments in the familiar surroundings of early Victorian country life. These and contemporary examples show that Darwin's insights apply with equal force to issues that are shaping your life and the lives of your descendants.

Darwin stressed the complex relations that exist among forms of life far removed in the scale of evolution from one another and between those life forms and the environment. He pointed out the synergy that exists between an ecosystem and evolution. As the environment changes, whatever might confer a selective advantage changes; so as we humans change the world, it changes us. And if we don't take care, it could be the end of us.

Providence, R.I.

February 3, 2017

ACKNOWLEDGMENTS

've drawn on the research of my colleagues over the course of more than fifty years as well as on people I have met only through their books and papers. I cannot tell Elizabeth Bates, Arend Bouhuys, Franklin Cooper, Edmund Crelin, Dennis Klatt, Karen Landahl, Bill Laughlin, Alvin Liberman, Molly Mack, or Kenneth Stevens that I valued their thoughts and work. The speech-research groups at Haskins Laboratories, MIT Lincoln Laboratories, and Air Force Cambridge Laboratories and MIT's Research Laboratory of Electronics have all contributed in some way to this book. Noam Chomsky at the start and Morris Halle were both part of the story, though our paths have since diverged. I can still express my appreciation to my son Daniel Lieberman for his insights on human evolution, John Shea for his reappraisal of the significance of the archaeological record of stone-tool fabrication and use, David Gokhman and his colleagues for their genetic insights on of the evolution of human vocal anatomy, and Robert McCarthy for perfecting the reconstruction of the vocal tracts of extinct hominins. Sheila Blumstein, Joseph Friedman, Geoffrey Laitman, Angie Morey, Emily Pickett, Joan Sereno, and scores of students at Brown University were colleagues in research. CARTA (Center for Academic Research and Training in Antrthropogeny) meetings over the course of the last two decades pointed out both answers and new questions. Marcia Lieberman was my constant companion in research at strange locations in Nepal and the Himalayas, such as Everest Base Camp, Lo Manthang, and Inner Dolpo. She has challenged me in the more than fifty years that we've been together and spent days unscrambling my prose. Eric Schwartz, Robert Fellman, and the editorial staff at Columbia University Press did their best. But the infelicities and errors that you may find are mine.

THE THEORY
THAT CHANGED
EVERYTHING

1

STRAWBERRIES

Doubtless God could have made a better berry, but doubtless God never did.

—Izaak Walton, *Compleat Angler, or the Contemplative Man's Recreation*, 1653

Concerning creating a better berry, Charles Darwin, in the first edition of *On the Origin of Species* (1859), had a different view:

> Gardeners picked out individual plants with slightly larger, earlier, or better fruit, and raised seedlings from them, and again picked out the best seedlings and bred from them, then, there appeared (aided by some crossing with distinct species) those many admirable varieties of the strawberry which have been raised during the last thirty or forty years.[1]

Some plants yielded bigger strawberries, while strawberries on other plants were sweeter. These plants were selected, sometimes grafted, and cultivated. Some plants had berries that weren't sweet at all and were discarded. Strawberries, along with pears, potatoes, roses, dahlias, and other plants, were "perfected" by human beings selecting and cultivating plants that had desirable characteristics. Variation is the normal state of things for all living organisms—for every species, vegetable and animal, including "identical" twins. Darwin had no knowledge of the biological mechanisms that result in variation, but, as he put it, "No one supposes that all the individuals of the same species are cast in the same mould."[2] Over the course of thousands of years, humans selected and bred plants and animals that had desirable variations. They also "crossed" different species, producing genetically modified organisms. Darwin didn't use the term GMO, but they are not new—you would have to forage in the forests and meadows to find any food that has not been genetically modified. When gardeners graft a cutting from one plant onto another, they produce a GMO.

In what follows, I will describe a variety of examples that demonstrate the complex interplay between natural selection and the environment, between the pace of evolution and its relation to human wants and behavior.

* * *

By 1859, agriculturalists had produced large, juicy strawberries that differed markedly from tiny, wild strawberries. Over time, selection and grafting had also produced more colorful flowers, repeated selection had produced sheep that yielded more wool, and dairy farmers had bred cows that produced more milk. Despite his recurring bouts of ill health, Darwin continued to collect data. He arranged to meet experts on the cultivation of lilies and ferns. He sought out gardeners, poultry breeders, cattle and horse breeders, and dog fanciers, and he quizzed coachmen concerning horses. He printed and mailed out questionnaires on breeding horses, cattle, sheep, and other animals. The first chapter of *On the Origin of Species* gives dozens of examples of "artificial selection." The key was "man's power of accumulative selection: nature gives successive variations; man adds them up in certain directions useful to him."[3]

On the Origin of Species is a long, occasionally wandering chat with an exuberant scholar who is pouring out a stream of ideas, an enthusiastic writer whose store of knowledge reflects his unusual experiences in odd locales and his intense, wide-ranging scholarship. "The principle of selection I find distinctly in an ancient Chinese encyclopedia. Explicit rules are laid down by some of the Roman classical writers." He studied "treatises published on some of our old cultivated plants, as on the hyacinth, potato, even the dahlia, &c." He reveals with delight obscure facts plucked from singularly rare tracts. Devoting eight pages to a discussion of the anatomy of pigeons, he noted that "in the time of the Romans, as we hear from Pliny, immense prices were paid for pigeons."[4]

THE *BEAGLE*

Toward the end of the short autobiography and collection of his father's letters that Darwin's son Francis edited is Charles's letter to Captain Robert FitzRoy, the captain of HMS *Beagle*. Charles wrote, "What a glorious day the 4th of November will be to me— my second life will then commence, and it shall be a birthday for the rest of my life."

The *Beagle* did not sail until December 27, 1831, but as Darwin concluded in the brief autobiography that he wrote for his children and grandchildren: "The voyage of the *Beagle* has been by far the most important event in my life, and has determined my whole career." Fresh out of Cambridge University, Charles Darwin was chosen by Captain FitzRoy as his gentleman companion, someone with whom he could dine and freely converse on the four-year voyage. Between 1831 and 1835, the *Beagle* first circumnavigated South

America, then sailed across the Pacific and around Africa to return to England. What Charles Darwin witnessed at sea and on land (where he spent more time over those four years) transcended anything offered today in an "adventure travel" brochure and was completely outside the social and conceptual framework of early Victorian England: for example, he wrote, "we see the value set on animals even by the barbarians of Tierra del Fuego, by their killing and devouring their old women, in times of dearth, as of less value than their dogs."[5] The openness of Darwin's mind was nonetheless apparent. The "barbarians" to Darwin were as human as his English shipmates.

Three Fuegians were aboard the *Beagle* throughout the voyage down the eastern coast of South America to Tierra del Fuego—two young men, York Minster and Jemmy Bucket, and Fuegia Basket, a girl of twelve or thirteen years. The Fuegians' names had been coined and bestowed on them five years earlier during the *Beagle's* previous voyage to Tierra del Fuego. Captain FitzRoy had "collected" York Minster and Fuegia Basket and had bought Jemmy Bucket for about fourteen dollars.[6] The Fuegians were homeward bound to fulfill FitzRoy's intention to found a Christian community and mission station. During their stay in England, they had become minor celebrities. They had learned Victorian table manners and proper dress, and they could speak English to varying degrees. They were evasive when questioned about their beliefs, since they probably realized that whatever gods they worshiped or beliefs they held would be regarded as childish, barbaric, or both. The three Fuegians, nonetheless, displayed the individual foibles, temperaments, and attitudes that identified them as unmistakably human:

> We had now on board, York Minster, Jemmy Button (whose name expresses his purchase-money [a pearl button]), and Fuegia Basket.

York Minster was a full-grown, short thick, powerful man; his disposition was reserved, taciturn, morose, and when excited violently
passionate; his affections were very strong towards a few friends on
board; his intellect good. Jemmy Button was a universal favorite, but
likewise passionate; the expression of his face at once showed his nice
disposition. He was merry and often laughed, and was remarkably
sympathetic with anyone in pain. . . . He was of a patriotic disposition; and liked to praise his own tribe and country, in which he said
there were "plenty of trees," and he abused other tribes: he stoutly
declared that there was no Devil in his land. Jemmy was short, thick
and fat, but vain of his personal appearance; he used to wear gloves
and his hair was neatly cut, and he was distressed if his well-polished
shoes were dirtied. He was fond of admiring himself in a looking-
glass. . . . Lastly, Fuegia Basket was a nice, modest, reserved young
girl, with a rather pleasing but sometimes sullen expression and very
quick in learning anything, especially languages. This she showed
by picking up some Portuguese and Spanish, when left on shore for
only a short time at Rio de Janeiro and Monte Video, and in her
knowledge of English.[7]

Darwin devoted an entire chapter of *The Voyage of the* Beagle to
describe the different indigenous inhabitants of Tierra del Fuego
and their reception of Jemmy, Fuegia, and York. He describes his
first contact in close detail:

Their only garment consists of a mantle made of guanaco skin, with
the wool outside; this they wear just thrown over their shoulders,
leaving their persons as often exposed as covered. Their skin is of a
dirty coppery red colour. The old man had a fillet of white feathers
tied round his head, which partly confined his black, coarse, and
entangled hair. His face was crossed by two broad traverse bars; one,
painted bright red, reached from ear to ear and included the upper

lip; the other, white like to chalk, extended above and parallel to the first, so that even his eyelids were thus colored.

On another day, he encountered a different tribe:

> These Fuegians in the canoe were quite naked, and even one full grown woman was absolutely so. It was raining heavily, and the fresh water, together with the spray, trickled down her body. In another harbor not far distant, a woman, who was suckling a recently-born child, came on alongside the vessel [one of the *Beagle*'s boats], and remained there out of mere curiosity, whilst the sleet fell and thawed on her naked bosom, and on the skin of her naked baby! . . .
>
> At night, five or six human beings, naked and scarcely protected from the wind and rain of this tempestuous climate sleep on the wet ground coiled up like animals. Whenever it is low water, winter or summer, night or day, they must rise to pick shellfish from the rocks; and the women must either dive to collect sea-eggs, or sit patiently in their canoes, and with a baited hairline without any hook, jerk out little fish. If a seal is killed, or the floating carcass of a putrid whale is discovered it is a feast.[8]

Jemmy Button, Fuegia Basket, and York Minster, on returning home, dispensed with their European clothes. York took Fuegia as his wife and left to join his own tribe, but a strong bond had formed between Jemmy and Captain FitzRoy and the *Beagle*'s crew:

> On the 5th of March, we anchored in the cove at Woolya, but saw not a soul there . . . soon a canoe with a flag flying was seen approaching, with one of the men in it washing the paint off his face. The man was poor Jemmy,—now a thin haggard savage, with long distended hair, and naked, except a bit of blanket round his waist. We did not recognize him till he was close to us; for he was ashamed of himself and

turned his back to the ship. We had left him plump, fat, clean, and well dressed;—I never saw so complete and grievous a change. As soon however as he was clothed, and the first flurry was over, things wore a good appearance. He dined with Captain Fitz Roy and ate his dinner as tidily as formerly. . . . With his usual good feeling, he brought two beautiful otter-skins for two of his best friends, and some spear-heads and arrows made with his own hands for the Captain. On the departure of the *Beagle*, Jemmy who had returned to shore with his new wife, lit . . . a signal fire, and the smoke curled up, bidding us a last and long farewell as the ship stood on her course into the open sea.[9]

Tierra del Fuego made it clear to Darwin that all people, of whatever color, are brothers and sisters, shaped by culture and circumstance.

When he first boarded the *Beagle*, Darwin had no doubt that God had created the world and all forms of life, including humans, and he accepted the biblical account in the book of Genesis. The voyage of the *Beagle* was the transformative event in Darwin's life, priming his mind for his transformation of biological science. In the autobiography that he wrote in the closing year of his life, Darwin recounted his vivid memories of his voyage on the *Beagle*.

The glories of the vegetation of the Tropics rise before my mind at the present time more vividly than anything else; though the sense of sublimity, which the great deserts of Patagonia and the forest-clad mountains of Tierra del Fuego excited in me, has left an indelible impression in my mind. The sight of a naked savage in his native land is an event which can never be forgotten. Many of my excursions on horseback through wild countries, or in the boats, some of which lasted several weeks, were deeply interesting. . . . I also reflect with high satisfaction on some of my scientific work, such as solving the

problem of coral islands. . . . Nor must I pass over the singular rela-
tions of the animals and plants inhabiting the several islands of the
Galapagos archipelago, and all of them to the inhabitants of South
America.

The publication of *The Voyage of the* Beagle also was Charles Darwin's entrée into the world of science: "As far as I can judge of
myself, I worked to the upmost during the voyage from the mere
pleasure of investigation, and my strong desire to add a few facts to
the great mass of facts in Natural science. But I was also ambitious
to take a fair place among scientific men, whether more ambitious
or less so than most of my fellow workers I can form no opinion." By
1837, two years after the *Beagle*'s return to England, Darwin had
embarked on another voyage, in the depths of his mind, toward the
theory of evolution by means of natural selection.

DARWIN'S FINCHES

The Galapagos archipelago is about six hundred miles from the
western coast of South America. It has ten main islands, none of
which is large, and several other smaller islands. Adventure-travel
companies now schedule tours on cruise ships large and small, at
prices ranging from expensive to astronomical. The iguanas, giant
tortoises, and finches are the star attractions. The tortoises are usually featured in the brochures and websites, probably because they
are easier to photograph and stranger than the birds. Snorkeling,
swimming, and excursions on inflatable Zodiac boats, on which
everyone must wear a serious-looking life jacket, enliven the trip.
Lectures on Darwin and evolution are practically mandatory. The
talks almost always focus on "Darwin's finches." The oft-told story
is that Darwin's eureka moment occurred when he examined the

finches that he had collected on the different islands of the Galapagos.

Darwin's initial focus was the geology of the Galapagos, but he was impressed by the giant tortoises. He considered the iguanas hideous and was amazed to find animals and birds that were so tame. He could walk up to a bird and nudge it with the barrel of the gun he carried without disturbing it. He tugged on the tail of an iguana, and it didn't run away. Darwin and his shipmates collected specimens of the flora and fauna. The tortoises were also brought aboard the *Beagle* alive as a source of fresh meat on the voyage to Tahiti.

Darwin collected many birds, classifying most of them by beak shape—whether their beaks were long and tapered to a point or short and stubby. He noted that the mockingbirds were different from island to island. However, he carelessly failed to keep track of which island which of the birds had lived on. The Galapagos, in fact, were not actually where Darwin had his epiphany about evolution and natural selection.

The birds, tortoises, and iguanas of the Galapagos nonetheless have a special place in the history of science. The distinctions that mark living species and their varieties are fuzzy. Although all horses are members of the same species, Shetland ponies and the large Clydesdale cart horses seen in beer commercials evince major differences. Tiny Tibetan terriers and lumbering St. Bernards are different varieties of dogs. Moreover, being able to mate and produce fertile offspring isn't the distinguishing criterion for speciation. Coyotes and wolves can mate, producing coywolves, which are now common in the eastern United States. On the long voyage back to England, Darwin began to realize that the boundary between a species and its varieties was fluid. "When I see these islands in sight of each other, & possessed of but a scanty stock of animals, tenanted by these birds, but slightly differing in structure & filling the same place in Nature, I must suspect they are only varieties . . . such facts undermine the stability of Species."[10]

By 1859, more than twenty years after the *Beagle*'s return to England, this position had become to Darwin a certainty—the boundary between species and variety is arbitrary because a species can gradually evolve to become a "new" species:

> I look at the term species, as one arbitrarily given for the sake of convenience to a set of individuals closely resembling each other, and that it does not essentially differ from the term variety, which is given to less distinct and more fluctuating forms. The term variety, again, in comparison with mere individual differences, is also applied arbitrarily, and for mere convenience sake.[11]

The process by which he reached that conclusion and a theory of evolution that could be backed up with facts took decades. In 1837, after his return to England, the specimens collected on the *Beagle*'s voyage were examined and classified. The specimens collected by Captain FitzRoy, the captain's servant, and the *Beagle*'s first lieutenant were all labeled to note which island they came from. The zoologists in England pronounced the birds, tortoises, and iguanas from each island to be different species.

However, it became apparent to Darwin that slight, graded distinctions differentiated the hypothetical species of Galapagos birds, tortoises, and iguanas that lived on different islands. He had also collected the remains of extinct species. Though the fossils that he had collected often were quite different from any living animals, some seemed to be related to living species. The idea of the "transmutation" of species—evolution—slowly formed in his mind. And he recorded these thoughts in a notebook, the first of five. During the two decades before the publication of *On the Origin of Species*, Darwin worked at a furious pace. His narrative of the *Beagle*'s voyage was his entrance into the ranks of "natural philosophers." He presented papers at the meetings of both the Geological and Zoological Societies. He published three books describing the geology of coral reefs, volcanic islands, and South America, wrote three monographs on barnacles and one on fossil barnacles, and edited five books on the zoology he observed during the *Beagle*'s voyage. All the while he was engaged in working his way toward explaining the transmutation of species, continually gathering data, and seeking out anyone who had relevant knowledge: gardening, animal husbandry, or pigeon breeding, for example.

While doing all this, after an exceedingly brief courtship, he married his cousin, Emma Wedgwood, whom he had known since childhood. His 1844 "essay" is an abbreviated, chapter-by-chapter version of *On the Origin of Species*. Darwin left instructions to Emma to have it published if he died before completing the final version.

NATURAL SELECTION

Biology owes to Darwin the premise that evolution is not directed by God or by any "laws of nature":

> Any variation, however slight and from whatever cause proceeding, if it be in any degree profitable to an individual of any species, in its infinitely complex relations to other organic beings and to external nature, will tend to be inherited by its offspring. The offspring, also, will have a better chance of surviving. . . . I have called this principle, by which each slight variation, if useful, is preserved by the term of Natural Selection.[12]

Natural selection acting on variation was to Darwin the primary engine for the "transmutation" of species. The direct action of the environment could, as Charles's grandfather Erasmus Darwin had proposed in *Zoonomia; or, The Laws of Organic Life*, published in 1794 and 1796, also play a role in forming "new" species from "old" ones, but natural selection was the prime agent in the origin of species.

Moreover, there is no clear criterion that signals the creation of a new species. In one of his books on barnacles, *Living Cirripedia* (1844–1845), Darwin documented the range of variation within a species, reinforcing his conclusion that the distinction between a variety and a species was arbitrary. Throughout *On the Origin of Species*, Darwin stressed the gradual nature of evolution. However, although Darwin did not realize this, much more rapid changes can occur. In the Galapagos, the ocean currents of the Pacific—El Niño, the warm phase, and La Niña, the cool phase—affect the growth of the nuts that the finches feed on. Had Darwin spent ten or twenty years in the Galapagos, he would have seen that he was wrong. Within a few generations, natural selection can favor the proliferation of finches that have short, strong beaks that can break open tough-skinned nuts or else longer beaks better adapted to feeding on softer nuts.[13] It is evident that the pace of natural selection on humans can also be rapid.

THE PACE OF NATURAL SELECTION

Experiences like Darwin's on the voyage of the *Beagle* clear the mind. The ambassador to the United States from Nepal once told me that his country belonged to "the fourth world." He knew his country well, but even he was not aware of the extent to which his remark was true. My wife, Marcia, and I trekked across much of Nepal, making journeys on foot into its most remote, isolated areas, sequestered behind high mountains, accessible only by strenuous and sometimes risky footpaths, giving us direct experience of places and lives in Nepal the ambassador himself had never directly witnessed.

In northwestern Nepal, in the area known as Inner Dolpo, there was nothing to indicate that we were still in Nepal or, for that matter, exactly where we were. The Bon-Po religion that preceded Buddhism survives in isolated regions such as Inner Dolpo, and we were walking from village to village to see if we could find any traces of ancient Bon-Po art in the few monasteries.

Not only was there no road; there was no policeman or official to be seen, no post office, school, medical clinic, or telephone. Not even so much as a note posted on a tree or a sign on a house to indicate the name of where we were. The two of us traveled for six weeks, helped by a crew of porters and guides. There, and in other parts of backcountry Nepal, after long hours together, we learned much about their lives. I treated scrapes and wounds by washing them, something they had not thought of doing. I also disinfected cuts and applied bandages from our medical kit. This kind of treatment was unknown to them.

As we walked, Marcia found herself conducting a sort of moving school for long hours and many days. The Sherpa guides acted as interpreters, and a few porters could speak some English. Questions were posed forth: Do you have porters in America? Is the Earth really round? Are Lenin and Marx alive? (An election was coming up in which the United Marxist-Leninist Party was fielding candidates.) One man said his father had cut his leg on a rock and a week later was dead. When Marcia asked how many had a father still alive, only two or three did. We discovered that they were unaware of the existence of, for example, bacteria. They had never heard of the germ theory of disease.

In our time, we easily forget that the germ theory of disease was not taken seriously by "Western" medicine until Louis Pasteur's work in the 1860s. The three Brontë sisters all died of tuberculosis in the late 1840s. Charles Darwin, like the physicians of his time, thought that bad air played a role in transmitting tuberculosis. Childbirth was more dangerous in a hospital than at home, owing to physicians not washing their hands after examining sick patients or conducting autopsies. In this setting, natural selection shaped human populations, often rapidly.

The Black Death, the second pandemic in recorded history, swept through Asia, arrived in Sicily in 1347, and in three years reached most of Europe. Precise data are not available, but over the next five years as much as 60 percent of Europe's population died from the bubonic and pneumonic forms of the plague, caused by the bacteria *Yersinia pestis*. The bubonic form, transmitted by bites from fleas that had fed on rats that carried the disease, created swellings (buboes) in the armpits, groin, and neck. The pneumonic form of the plague attacked the lungs and spread to other people through coughing and sneezing. When the bacteria entered the blood, there was little hope of survival.

Analyses of DNA from Bronze Age teeth show the presence of *Yersinia pestis* in Europe and Asia as far back as five thousand years ago.[14] Mutations in the bacteria increased its virulence and allowed it to be transmitted by fleas. The first pandemic, the Plague of Justinian, occurred in AD 541, in Rome. The fourteenth-century European Black Plague continued with intermittent outbreaks until the eighteenth century. The Great Plague of London that Daniel Defoe described in his *Journal of the Plague Year* started in the spring of 1665 and killed more than one hundred thousand people. It ended in 1666, when the Great Fire of London burned down 13,000 houses, perhaps killing off the rats. A third pandemic started in China in the 1850s and spread around the world. No effective cure was available: the key to survival was a robust immune system. If your ancestors were European or Asian, you most likely possess a genetically transmitted degree of immunity to *Yersinia pestis* owing to these "selective sweeps."

Natural selection operates on individuals. A horticulturist might plant the seeds of only those strawberry plants that produce juicier berries. In the state of nature, if some heritable characteristic leads to some individual's offspring having a better chance of surviving, then in time that characteristic may *sweep* through the population. And in many circumstances, natural selection can operate swiftly. If every being that does not have a particular heritable variation dies, the selective sweep can be rapid. That was the case for the plague and other endemic Eurasian diseases.

When Europeans arrived in the Americas, they brought their endemic diseases with them, which virtually wiped out the native populations. In 1541, the Spanish conquistador Francisco de Orellana, who had made his name in the conquest of Ecuador and Peru, started a journey that eventually took him and fifty-seven Spanish "companions" down the length of the Amazon. Their journey was recorded by Friar Gaspar de Carvajal. They had started on an ill-fated expedition with the brother of Francisco Pizarro, who had conquered the Inca Empire. The expedition was running out of supplies, and Pizarro dispatched them down a river to search for food. However, it was impossible to return against the current, and the land route was forbidding.

The most viable option was to run with the current and continue descending the river. Orellana established friendly relations with one native community, and the Spanish were able to cut timber, smelt iron for nails, and build two large boats, "brigantines." The river flowed into the Amazon. Food was a constant problem, which the conquistadors often solved by looting villages—leading to almost continual warfare. Orellana and his surviving companions, including Friar Carvajal, who lost an eye to an arrow in one battle, reached the Atlantic and safety in the Spanish settlement of Nueva

Cadiz on the island of Cubagua, off the coast of Venezuela. Carvajal describes towns stretching more than sixty miles along the banks of the river and densely populated villages. His account has been disputed because no trace of these settlements and their inhabitants was evident when the Portuguese arrived in Brazil sixty years later. Recent mapping of the Amazon using ground-penetrating radar is beginning to reveal stone monuments and other traces of thriving complex civilizations.

But it was contact with Orellana and his men that most likely transmitted the endemic European diseases that devastated the native population. Spanish accounts of depopulated provinces in the Inca Empire suggest that 25 to 90 percent of the indigenous population died in the sixteenth century; the population of Peru did not return to its Inca level until the twentieth century.[15] In the Americas, contact with Europeans in the centuries that followed transmitted smallpox, measles, influenza, and the bubonic plague, killing hundreds of thousands of people in North America alone.

YOUR SUPERMARKET'S SHELVES

As is the case for Darwin's finches, the menu can drive natural selection in comparatively short timeframes. You don't need to visit the Galapagos to see the evidence; it's on the shelves of your local supermarket. If you can drink a glass of milk without unpleasant gastrointestinal consequences, you owe that capability to natural selection acting on your ancestors, who would have lived in Northern Europe, the Kenyan highlands, or other places where herds of cows, sheep, goats, or yaks were domesticated and milked.

Selective sweeps have brought new sources of food to the menu recently in the timescale of human evolution. We humans diverged from Neanderthals somewhere between 700,000 and 500,000 years ago. In contrast, cows, goats, and sheep were first domesticated ten to seven thousand years ago, and milk from domesticated animals was added to the menu no earlier than about seven thousand years ago. After weaning, older children and adults have difficulty digesting milk products unless their ancestors lived in a culture in which natural selection conferred adult lactose tolerance because milk products were present.

In Kenya, natural selection for adult lactose tolerance developed only three thousand years ago, when cattle were introduced from Ethiopia.[16] A selective sweep occurred: the added source of food conferred a selective advantage to adults who possessed this heritable capability; more of their children survived and then had children who themselves survived. The gene frequencies of the local population changed as increasingly more people and their children who possessed adult lactose tolerance survived. Since not everyone living today has ancestors who herded and milked cows, sheep, yaks, or goats, you'll find lactose-free food products on supermarket shelves. Other instances of natural selection and selective sweeps that opened up different sources of food are coming to light. Most people wouldn't reach Medicare age if they were ate a traditional Inuit diet. The Inuit or Eskimos (the word "Eskimo" is not a pejorative to the people so described) crossed what's now the Bering Strait on a land bridge. The Inuit ate what was available: seals, seals, more seals, fish—and nothing else. They can survive into old age without blocked arteries and vitamin deficiencies owing to natural selection and a selective sweep that yielded unique metabolic capabilities.[17]

INNER OVENS AND BREATHING

Natural selection has acted to adapt humans to life in different ecosystems. People whose ancestors lived for long periods of time in the Arctic, such as the Inuit, have "inner ovens"—the maternally transmitted DNA in these Arctic peoples' mitochondria code for a higher metabolism, which keeps their host cells warmer. Selective sweeps occurred in people that lived in Siberia, including the ones who left for the Americas. But this genetic variation is present in individuals whose ancestors never lived in the Arctic. Natural selection acted on mitochondrial variations that are present in the general human population to favor surviving in the Arctic. I'm usually comfortable taking a walk on a late fall day wearing a T-shirt, but my wife wears two fleece jackets. Outdoors in December and January, she's encased in down-filled parkas. My mitochondria are burning a bit hotter, though it's unlikely that any of my ancestors spent any time in the Arctic.

Switzerland may seem an unlikely place to search for heritable variations other than tolerance for drinking milk and eating cheese. Starting from a mountain hut above Zermatt, Switzerland, I climbed the Matterhorn in the 1970s. It took a bit over four hours, the usual time for the 1,200-meter climb on the "normal" route used by 99 percent of climbers ascending from Zermatt. The speed record for the normal route then was about ninety minutes. The remaining 1 percent of climbers usually ascend on the demanding Zmuttgrat route. In 2015, Dani Arnold climbed the incomparably difficult and dangerous North Face in one hour, forty-six minutes. The few superclimbers who have risked the North Face have taken eight to ten hours. Apart from his skill, willingness to take risks, and luck, how was Dani Arnold able to set this record time?

In the early decades of the nineteenth century, it became evident that life depends on oxygen being transferred from the air that we breathe into our blood. If any divine plan had created human beings, you might expect to find that we all share the same efficient system, but that isn't the case. Respiratory physiologists have developed a simple technique that determines the efficiency of the mechanism that transfers oxygen from your lungs into your bloodstream. Subjects vigorously pedal a bicycle or run on a treadmill while the amount of air they are breathing and the oxygen level in their bloodstream is measured. Surprisingly, some people must inhale ten times as much air as others to transfer the same amount of oxygen into their blood.[18] It generally doesn't matter whether you live in Amsterdam, parts of which are below sea level, or in Denver, Colorado, at an altitude of 4,500 feet. This aspect of respiratory efficiency instead is determined by your genetic endowment, and people vary, despite hundreds of millions of years of evolution, which you might have assumed would result in optimal respiratory efficiency for everyone.

Nevertheless, respiratory efficiency has been shaped by natural selection. The only people who as a group have better respiratory efficiency are those whose ancestors lived for thousands of years at extreme altitudes. Natural selection acted to enhance the survival of individuals and their children whose oxygen-transfer mechanism was able to cope better with thinner air. The Sherpa guides and porters who make it possible for climbers to reach Mount Everest's summit emigrated from the high Tibetan plateau about three hundred years ago. As my wife, Marcia, observed during many of our treks in Nepal, Sherpa comfort food is boiled potatoes and yak cheese garnished with pepper paste. At Everest Base Camp, I once saw a Sherpa starting the 2,400-foot climb up to Camp Two carrying a fifty-pound sack of potatoes.

Respiratory physiologists running standard exercise tests found that people of Tibetan ancestry were almost all at the high end of efficiency. Independent genetic studies confirm that their DNA includes genes that optimize oxygen transfer from their lungs to their bloodstream. Interestingly, natural selection on people living in the Andean mountains of South America came up with a different solution—very large lungs.[19] Apart from his skill, Dani Arnold, who has no known Tibetan or Andean ancestors, must have luckily inherited an exceptionally efficient respiratory capability.

EVEREST

Studying climbers ascending Mount Everest added surprising insights on the nature and evolution of the neural bases of human cognition. After arriving in Kathmandu, the site of Nepal's only international airport, virtually all Everest climbers plan on taking at least two months to acclimatize to the extreme altitude. After they fly to the airstrip at Lukla in the Everest region of Nepal, climbers must gradually ascend, staying several nights in "lodges" along the way, reaching Everest Base Camp, at 17,500 feet, after about ten days. Meanwhile, the climbing route up Everest is being prepared by the "Ice Doctors," an intrepid group of Sherpas who hold the key to Everest—getting through the Ice Fall, a slope that threads through a river of glacial ice flowing down 2,400 feet. They haul up long aluminum ladders and secure them to the ice to form bridges that span crevasses—deep chasms in the ice. Crevasses shift and reform almost daily: the river of ice flows four or five feet per day. As the ice flows, it spawns another hazard, seracs: blocks of ice that can be as large as a house and can dislodge and fall. Avalanches of snow and ice can crash down without warning. At Base Camp, which is usually protected from avalanches, you can hear several avalanches on most days. On April 18, 2014, sixteen Sherpas died when a magnitude 7.8 earthquake triggered an avalanche that thundered down the Ice Fall. Base Camp was swept by an avalanche from nearby Mt. Pumori.

In a normal year, after the Ice Doctors complete their work, Sherpa porters must carry tents, stoves, fuel, oxygen, and miles of rope up to a series of four camps. Most of the climbers continue to wait at Base Camp. The waiting time is useful because physiological studies show that it takes at least fourteen days to achieve 50 percent acclimatization. The ascent of Everest then begins. The trip through the Ice Fall is the only the start of a series of climbs to higher camps and the summit. After threading through the Ice Fall, climbers reach Camp One at 20,000 feet and usually go directly up to Camp Two, at 21,300 feet. Camp Two is the most pleasant resting place above Base Camp. It is relatively protected from avalanches and stocked with provisions.

Meanwhile, Sherpas continue to carry up miles of rope, pounding hundreds of aluminum stakes into packed snow and inserting long ice screws into icy sections. Almost no one would be able to climb Everest without the Sherpas. When their work is complete, a rope anchored to the mountain stretches from Camp Two to Camp Three at 24,000 feet, and on to Camp Four at 26,000 feet, at the South Col of Everest. When the winds on Everest die down and the weather for the next few days seems to be stable, the mass of climbers who have remained at Base Camp ascend to Camp Two. When they reach the fixed ropes, they each attach a rope clamp that they slide up the rope as they slowly ascend. From a distance, the climbers in their red, yellow, or blue down parkas resemble a brightly colored caterpillar crawling up the slope. The slowest climber sets the pace. The final stage, above the fixed ropes from Camp Four to the summit at 29,035 feet and back, can take up to nineteen hours.

In 1993, I was at Everest Base Camp with a two-person Brown University research team to study the effects of low oxygen levels on climbers. A previous independent study had pointed to permanent cognitive deficits in world-class climbers who had been exposed to extreme altitudes. Moreover, dozens of aviation disasters had been attributed to pilots making inexplicable errors when oxygen systems malfunctioned, gradually decreasing the oxygen level within the plane. Our NASA-funded project was aimed at developing a speech-analysis system that would allow aircrews to be monitored remotely for oxygen deficits by detecting motor-control problems. It would have been unethical to expose subjects to low oxygen levels deliberately. Therefore our volunteer subjects were recruited from climbers at Everest Base Camp who had already decided to ascend.

Our 1993 study and our subsequent 2001–2008 Everest studies revealed behavioral problems deriving from sustained low blood-oxygen levels. As our subjects ascended Everest, speech motor control slowly deteriorated. Sentence comprehension became impaired, and cognitive flexibility, the ability to change the direction of one's plans, deteriorated. In some instances, emotional control also lapsed.[20]

At Base Camp we first administered a battery of standard tests of cognition and indirectly monitored motor control by computer-implemented acoustic analysis of our subjects' speech. Talking involves precisely executing, timing, and coordinating tongue, lip, jaw, soft palate, and larynx movements; this is carried out by dozens of muscles. Musicians, acrobats, and athletes also learn and execute complex, rapid, coordinated motor acts, but everyone has to reach a similar level of skill in order to talk. Contrary to what you might assume, normal children are not able to talk at adult speeds or pre-ciseness until they're ten or twelve years old.[21]

We digitally recorded our subjects reading a list of words and sentences. Computerized acoustic speech analysis revealed effects that wouldn't be evident if you simply listened to the recordings. The procedures and analyses were similar to ones my colleagues and I had used in previous studies of Parkinson's disease, which degrades the basal ganglia, neural structures deep within the human brain. Complex motor tasks such as walking, talking, or tapping on a keyboard are controlled by neural circuits that link activity in different parts of the brain. The basal ganglia play a critical role in some of these circuits as well as in neural circuits that are involved in thinking.[22] We recorded each subject as he or she spoke a series of words and sentences. We also administered tests of the subjects' short-term verbal memory—the ability to remember words, sentence comprehension, and cognitive flexibility—and their ability to change the direction of a thought process or plan of action.

THE BASAL GANGLIA

In 1817, James Parkinson described the "shaking palsy," which started with tremor, muscle stiffness, and motor impairments. Every motor act becomes slower, walking becomes difficult, and talking slows down as vowels and pauses lengthen. Cognitive impairment often follows, manifested in difficulties with planning ahead, comprehending the meaning of sentences that have moderately complex syntax, and inflexibility—not being able to change the direction of a train of thought or plan of action.[23] Learning anything new can become difficult.

Parkinson's disease arises when the basal ganglia, neural structures deep within the human brain, are compromised. The immediate cause is thought to be insufficient dopamine, a neurotransmitter necessary for the functioning of the basal ganglia, but other factors may also enter into the disease process. A similar "syndrome" or set of behavioral deficits can occur when the basal ganglia are damaged by strokes or trauma.[24]

The motor deficits of Parkinson's disease generally are thought to reflect an impairment of the basal ganglia's "sequencing function." The basal ganglia act as a sort of neural switch that starts and stops the execution of the neural instructions of the submovements stored in the cortex (the outermost layer of the brain) that when linked together yield a complete action. For example, walking entails swinging your foot forward and then tilting it so that your heel strikes the ground. The basal ganglia activate and sequence these two submovements, which are stored in the motor cortex. In humans, the basal ganglia also form part of similar circuits that link to prefrontal and other regions of the cortex involved in thinking. The basal ganglia in these cognitive circuits again act as a switch, allowing us to change the direction of a thought process or plan of action.

The basal ganglia are metabolically active and need sufficient oxygen to function normally. The effective oxygen present in air decreases with air pressure at higher altitudes. For example, at 8,900 meters, Everest's summit, the air pressure has fallen to about one-third of that at sea level. Hence, as you ascend Everest, basal ganglia performance degrades, compromising motor control and cognitive flexibility, that is, performance on cognitive tasks entailing changing the direction of a thought process or plan of action. The human brain is a complex organ, and other neural capacities are involved in motor and cognitive acts. Therefore, individuals differ with respect to how precise or quick they are in performing motor acts as well as in their performance of cognitive tasks. Since the performance of Parkinson's patients on these tasks before the disease set in is almost always unknown, it is difficult to derive quantitative measures that would reflect basal ganglia impairment.

Our Everest studies allowed us to obtain baseline levels of performance at Base Camp, at 5,300 meters, before our subjects climbed to the higher camps, permitting an assessment of the behavioral deficits deriving from lower oxygen levels. MRIs performed elsewhere also showed that oxygen deficits could severely damage basal ganglia, sparing cortex.[25] We used radio links to run the same tests that had been administered at Base Camp and recorded the same set of spoken words and sentences when each climber first reached Camp Two and Camp Three. Radio reception was too poor at Camp Four to test subjects there.

Individual differences at altitude were apparent when our Everest data were analyzed. One subject, a twenty-year-old man, had normal performance when tested at Base Camp. Two days later at Camp Two, he had deteriorated to the point where his responses were similar to an advanced case of Parkinson's disease. His performance on cognitive tests showed that he was unable to adapt to changing circumstances. He "perseverated," thinking along the same lines even when circumstances demanded change. His speech had slowed and became distorted. Computer-implemented speech analysis showed that he had difficulty coordinating and timing the tongue, lip, jaw, and laryngeal motor gestures involved in talking. He insisted on following his original climbing plan, which called for ascending to Camp Three, at 24,000 feet. It was impossible to convince him to descend to Base Camp, and his friends at Camp Two did not intervene. He went up to Camp Three and fell to his death a few days later, unable to detach and then attach in proper sequence the carabiners that linked him to the anchor rope.

Our 2001–2008 Everest study had a different immediate objective. NASA had become aware of the danger that radiation in deep space posed to a manned mission to Mars. Focused radiation is used to treat cancer because radiation constricts arteries, reducing blood flow and oxygen. When mice are exposed to levels of radiation equivalent to that in deep space, their basal ganglia are damaged. On the long space voyage to Mars, crews might arrive debilitated by Parkinson's disease. Radiation also might damage another subcortical brain structure, the hippocampus, inducing Alzheimer's dementia. Absent some means for shielding crews from radiation, it is doubtful that humans will ever reach Mars.

The basal ganglia first appeared in froglike animals hundreds of millions of years ago. In frogs and lizards they control motor activity and also are involved in learning motor responses. In humans, they continue to play a part in motor control but have been "recycled" also to act as a neural switch in aspects of cognition such as

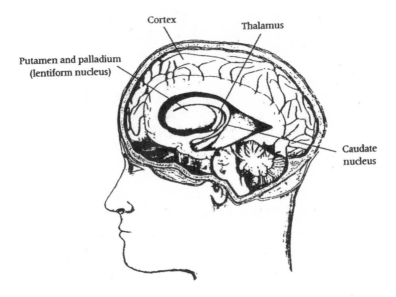

Cortex

Thalamus

Putamen and palladium
(lentiform nucleus)

Caudate
nucleus

FIGURE 1.1 Human basal ganglia

The basal ganglia, neural structures with a long evolutionary history, are buried deep within the human brain. The basal ganglia act in concert with other parts of the brain in regulating many aspects of human behavior, including walking, talking, and making decisions.

comprehending distinctions in meaning conveyed by syntax, changing the direction of a thought process, or responding to changing external circumstances. Other fatal disasters on Everest have involved experienced climbers not changing their climbing plans when the weather changed.

GOOD FENCES: THE ENVIRONMENT
ABOUT YOU

The synergy stressed by Charles Darwin between the environment and natural selection exists everywhere. The famous line in Robert

Frost's poem "Mending Walls," "good fences make good neighbors," brings to mind Darwin's observation of the consequences of a seemingly minor change to an ecosystem. Frost asks why fences make good neighbors, and the answer in the next line is that there are cows. As we change the world, it changes us. It's not only extreme environments such as Everest or the Arctic that determine which species will thrive or fail. Darwin was aware of the continual interplay between the environment and natural selection. Seemingly minor events could trigger a chain of events—anywhere.

In New England, some of the "commons" that grace small towns and a few cities still remain. At one time they served as pastures where everyone could graze their cattle or sheep. Though it is doubtful that you now would be allowed to lead a cow onto Boston Common, Boston still has a law on the books that states: "Anyone may let their sheep and cows graze in the public gardens or commons at any time except Sundays." In Hopkinton, Massachusetts, horses and cows are allowed on the common, but dogs are prohibited. The right to bring your sheep and cattle to the common, though hardly ever practiced, no longer exists in England.

The commons of New England were part of the culture that the early colonists brought with them from England, along with many place names—Boston, Acton, Shrewsbury, New London. However, in England by the beginning of the nineteenth century, the movement to "enclose" the commons—placing the land under the control and ownership of individuals—had accelerated. The process had started in the twelfth century and was complete by the end of the nineteenth. Throughout England, the "common" land on which members of a community could graze sheep or cattle was enclosed by a wall or fence or by virtual fences, that is, laws that forbade public use. Darwin observed that enclosing a village's common triggered a chain of environmental events. Cattle no longer grazed, and the vegetation changed. Insects that fed on particular plants vanished, the birds that fed on those insects changed, and so did the mammals that fed on the birds.[26] A simple fence, wall, or ordnance was all that was required to set natural selection in motion to effect systematic changes.

IN SHORT

As Charles Darwin realized, natural selection is a brilliantly simple process. Any heritable variation that fosters the survival of an individual and its offspring will be retained. What is retained depends on the ecosystem, which for humans brings in culture. No deity or mystical force directs the course of evolution. Its course involves the chance interactions of environment and the stock of heritable variations on which natural selection can act. Variation is the feedstock for natural selection and is the normal state rather than the exception. Darwin stressed the slow, gradual pace at which natural selection shapes living organisms, but he was wrong. If the selective advantage is extreme, allowing individuals who possess a favorable heritable variation to survive in stressful situations, natural selection can act rapidly. The effects of natural selection are evident on the shelves of your supermarket, in the resistance to or susceptibility to disease, and in such seemingly disparate endeavors as being able to climb Mount Everest or enjoy ice cream.

2

NO CATS, NO FLOWERS

The mind works in strange ways.

The myth is that an apple dropped from a tree and hit Isaac Newton's head, suddenly releasing the Universal Law of Gravitation from a deep corner of his mind. That's not exactly true—however, there *was* an apple.

On a late summer's day in 1666, Newton was relaxing in his mother's garden when he saw an apple fall from a tree. Many years later, according to the account recorded by William Stukeley, Newton said that he began to wonder:

> Why shd it not go sideways, or upwards? But constantly to the Earth's centre? Assuredly the reason is, that the Earth draws it. There must be a drawing power in matter. And the sum of the drawing power in the matter of the Earth must be in the Earth's centre, not in any side of the Earth. Therefore does this apple fall perpendicularly or towards the centre? If matter thus draws matter; it must be [in] proportion of its quantity. Therefore the apple draws the Earth, as well as the Earth draws the apple.[1]

Stukeley's account is preserved in the archives of the Royal Society. The ancient apple tree still lives on in Newton's mother's garden.

THE STRUGGLE FOR EXISTENCE

The particular copy of the book that triggered Charles Darwin's "eureka" moment hasn't been preserved. On an October day in 1838, he decided to relax and read for pleasure. Currently popular books to hand would have been Dickens's just-published novel *The Posthumous Papers of the Pickwick Club* or Mary Shelley's *Frankenstein: or the Modern Prometheus*. But for whatever reason, he decided to read Thomas Malthus's *An Essay on the Principle of Population*.

Almost fifty years later, in the short autobiography that he wrote for his children and grandchildren, Darwin described his apple moment. Since his return to England in 1837, he had been at work

> collecting all facts which bore in any way on the variation of animals and plants under domestication and nature. . . . I worked on true Baconian principles, and without any theory collected facts on a wholesale scale, more especially with respect to domesticated productions, by printed enquiries, by conversation with [skillful] breeders and gardeners, and by extensive reading. . . . I soon perceived that selection was the keystone of man's success in making useful races of animals and plants. But how selection could be applied to organisms living in a state of nature remained for some time a mystery to me.
>
> In October 1838, that is, fifteen months after I had begun my systematic enquiry, I happened to read for amusement "Malthus on Population," and being well prepared to appreciate the struggle for existence which everywhere goes on from long-continued observation of the habits of animals and plants, it at once struck me that under these circumstances favourable variations would tend to be preserved, and unfavourable ones to be destroyed. The result of this would be the formation of new species. Here then I had at last got a theory by which to work.[2]

Malthus had predicted wholesale disaster. Basic practices directed at improving health, such as removing open sewers full of human waste, had been gradually introduced throughout the eighteenth century. By the century's end, Europe's population had almost doubled, and Malthus foretold a runaway increase in population. Food would fall short. The starving masses would riot. Epidemics, riots, and war would destroy civilization. But the disaster never materialized. Food supplies increased faster than the growth of the European population. However, Malthus had provided Darwin with a theory—and with the phrase *"the struggle for existence."*

In *On the Origin of Species*, Darwin wrote that

> a struggle for existence inevitably follows from the high rate at which all organic beings tend to increase. . . . Hence, as more individuals are produced than can possibly survive, there must in every case be a struggle for existence, either one individual with another of the same species, or with the individuals of different species, or with the physical conditions of life. It is the doctrine of Malthus applied with manifold force to the whole animal and vegetable kingdoms; for in this case there can be no artificial increase of food, and no prudential restraint from marriage.[3]

Probably no phrase in the history of science has been so misinterpreted as this one. To most everyone, the "struggle for existence" translates into savage, ruthless competition—Alfred, Lord Tennyson's aphorism, "Nature red in tooth and claw." But that wasn't what Darwin meant. Tennyson was actually responding to Robert Chambers's 1844 book *The Vestiges of the Natural History of Creation*, which was the antithesis of the evolutionary theory presented in *On the Origin of Species*.

In what follows, I will discuss the medieval and Victorian theories concerning man's place in nature that *On the Origin of Species* disputed. I will also critique current simplistic models of the brain that date back to phrenology—bumps on your skull that are the "seats" of language, morality, recognizing faces, and most other capabilities. Darwin also introduced the biological mechanism of "recycling" (although he didn't use that term)—reusing existing organs to achieve new ends, a phenomenon that explains events such as the transition from aquatic to terrestrial life, which otherwise might be seen as evidence of God's will or some unspecified force directing the course of evolution.

VICTORIAN INTELLIGENT DESIGN AND PHRENOLOGY

The theory presented in *Vestiges*, like the creation myth of Genesis, accounted for both the formation of the universe and all living beings, including humans. The foundation of Chambers's theory was the medieval Great Chain of Being—in Latin, the *scala naturae* or stairway of nature, a hierarchy in which God stands at the top level, various forms of angels at lower levels, then the stars, the moon, and at lower steps humans, starting with kings, princes, knights, and finally commoners. Still farther down are animals, trees, plants, diamonds, rubies, gold, silver, metals, stones, etc.

Chambers departed from the account in Genesis, in which everything was created in six days and nights (seven, if you add the Sabbath). Chambers instead proposed that God had set up a master plan—a series of successive transformations toward "higher" forms of life. Life began with "lowly," simple forms that gradually progressed to higher, complex forms. Insects transformed into sea creatures, which were succeeded by mice, sundry animals, monkeys, and apes. Finally, the highest form of life was created—*Homo sapiens*. But not all humans were equal, and Europeans stood at the pinnacle.

Though God remained the creator of the universe and all forms of life, *Vestiges* was denounced by clerics because it deviated from the script that God had created everything over the course of a week. Nonetheless, *Vestiges* became a bestseller in Victorian England, running through twelve editions. Prince Albert even read *Vestiges* aloud to his wife, the young Queen Victoria.

Chambers also promoted phrenology, according to which innate mental "faculties," located in specific parts of the brain, governed virtually all aspects of human behavior. A "Faculty of Language" accounted for humans possessing language. An independent "Faculty of Music" accounted for that capability; a "Faculty of Numbers" accounted for arithmetic and the sense of time. The degree to which a person possessed one of these innate capabilities could be determined by measuring the size of a particular demarcated area of his or her skull. In practice, someone's sense of morality or mathematical ability could be assessed by the size of a bump on his or her head. Franz Josef Gall published the first phrenological map in 1796. The theory was elaborated by his disciple Johann Spruzheim. Phrenology was considered an established science in Edinburgh, where Darwin had briefly studied medicine.

George Combe in Edinburgh had elaborated on phrenological theory in 1828, in his self-help book *The Constitution of Man and Its Relation to External Objects*. Combe proposed "natural laws" that had to be obeyed to avoid divine punishment. Combe's book was a bestseller and suggested that if you followed its instructions, success awaited. About 350,000 copies of the successive editions of his book were sold between 1828 and 1900 in Great Britain, compared to fifty thousand copies of *On the Origin of Species*. Some faculties, such as the "Impulse to Propagate," were present in all animals. Other human faculties were present to a lesser degree in some animals. The faculty of language, however, was a unique human attribute.

The basic premises of *Vestiges* are alive and well today, fostered not only by proponents of "creation theory" and "intelligent design," who reject the concept of biological evolution, but also in mainline studies that purport to account for human attributes such as language, moral behavior, and art. The "modular" model of the human mind, popularized in Steven Pinker's books,[4] assumes that our brains are designed by computer nerds, who have put together a set of modules that each control one complex aspect of behavior, such as language or morality. The basic premise of Pinker's and similar modular theories can be traced back to phrenology—a particular part of your brain allows you to add up a series of numbers, another part of your brain allows you to talk, another part is your "organ" of fear, and so on. Supposedly we can recognize our cousin Allie's face because our brains contain a neural structure specifically designed to recognize human faces. The current research that will be discussed shows that this is clearly not the case.

THE WEB OF LIFE

It was clear to Darwin that the struggle for existence involves a complex web of interactions that transcend violence and ruthless competition. Unfortunately, Tennyson's characterization has become attached to the Darwinian theory of evolution. Almost every documentary film on evolution begins with images of huge snarling beasts pouncing on smaller ones for dinner. Some academic schools of economics take no-holds-barred competition as a lesson drawn from Darwin. The tenets of Social Darwinism and Ayn Rand's political philosophy misinterpret the struggle for existence to justify a "natural order" in which the world is for the strong.

Darwin realized that he would be misunderstood and his theory oversimplified. To Darwin, the struggle for existence instead entailed the complex interactions and dependencies that hold between living organisms and external events. And he hammered away on that point in *On the Origin of Species.*

> I should premise that I use the term Struggle for Existence in a large and metaphorical sense, including the dependence of one being on another, and including (which is more important) not only the life of an individual, but success in leaving progeny. Two canine animals in a time of dearth, may be truly said to struggle with each other which shall get food and live. But a plant on the edge of a desert is said to struggle for life against the drought, though more properly it should be said to be dependent on the moisture. . . . The mistletoe is dependent on the apple and a few other trees, but only in a far-fetched sense can be said to struggle with these trees, for if too many parasites grow on the same tree, it will languish and die. . . . In these several senses, which pass into each other, I use for convenience sake the general term of struggle for existence.[5]

Chapter 3 of *On the Origin of Species* is an argument by bombardment—facts and examples rain down on the reader. The struggle for existence instead involves long chains of seemingly unconnected events and actions—usually not entailing violent conflict or even competition in its ordinary sense. For example, after discussing topics ranging from the number of Scotch firs on heaths, insects, and the number of wild dogs and horses in Paraguay and insects and birds in Staffordshire, the reader learns that the absence of cats can mean no flowers:

> Plants and animals, most remote in the scale of nature, are bound together by a web of complex relations. . . . Many of our orchidaceous plants absolutely require the visits of moths to remove their pollen-masses and thus to fertilize them. I have, also, reason to believe that

humble-bees are indispensable for the fertilization of the heartsease (*Viola tricolor*), for other bees do not visit this flower. From experiments which I have tried, I have found that the visits of bees, if not indispensable, are at least beneficial to the fertilization of our clovers; but humble-bees alone visit the common red clover (*Trifolium pretense*), as other bees cannot reach the nectar. Hence I have very little doubt, that if the whole genus of humble-bees became extinct or very rare in England, the heartsease and red clover would become very rare, or wholly disappear. The number of humble-bees in any district depends in a great degree on the number of field-mice, which destroy their combs and nests. . . . Now the number of mice is largely dependent, as everyone knows, on the number of cats. . . . Hence it is quite credible that the presence of a feline animal in large numbers in a district might determine, through the intervention first of mice and then of bees, the frequency of certain flowers in that district!

Darwin is clear that

the structure of every organic being is related, in the most essential yet often hidden manner, to that of all other organic beings, with which it comes into competition for food or residence, or from which it has to escape, or on which it preys. This is obvious in the structure of the feet and talons of the tiger; and in that of the legs and claws of the parasite which clings to the hair on the tiger's body. But in the beautifully plumed seed of the dandelion . . . the relation seems at first confined to the elements of air and water. Yet the advantage of plumed seeds no doubt stands in the closest relation to the land being already thickly cloaked by other plants; so that the seeds may be widely distributed and fall on unoccupied ground.[6]

Darwin, however, was not a Pollyanna. In notebook E, one of the series of notebooks in which he recorded his thoughts on the transmutation of species, he glumly stated: "When two races of men meet, they act precisely like two species of animals,—they fight, eat each other, bring diseases to each other &c., But then comes the most deadly struggle, namely which has the best fitted organization or instincts (ie., intellect in man) to gain the day."[7]

It was clear to Darwin and virtually everyone else that some animals will fight and kill to eat. Other animals, such as bucks headbutting each other, fight to get control of mates. Jane Goodall, in her pioneering study *The Chimpanzees of Gombe*, documented scores of incidents of violence in Tanzania's Gombe National Park, including attacks in which chimpanzees were killed. For the most part, males were violent, but that was not always the case. Though present-day chimpanzees are not living embodiments of our common ancestor, our human propensity toward violence unfortunately seems to be an inherited attribute. Some pages of Goodall's book read like a war correspondent's journal. In the chapter on "Territoriality," Goodall wrote: "In 1972 we recognized the existence of a new community in Gombe—the Kahana community, which had previously been part of the large KK [Kasakela—Kakombe] study community. By the end of 1977, after an existence of five years, the Kahana community was no more."[8]

The chimpanzees living in different parts of the Gombe National Park in the early years of Goodall's research project mixed with one another, but by 1972 two distinct groups, a northern community based around the Kasakela and Kakombe valleys and a southern community based around the Kahana valley had formed. The Kahana chimpanzees were killed in a war. A video provided by Chris Boehm, one of Goodall's colleagues, shows chimpanzees silently and warily approaching the territorial boundary as they attempt to find a lone "enemy" chimpanzee they can mob and kill. The chimpanzees' faces are tense as they move cautiously in a single column toward the boundary. One adolescent male falls out of the column when they hear the calls of the enemy. An older chimpanzee places his arm around his shoulder and hugs him—and the adolescent returns to the column. But masked from view by the dense forest, neither they nor their adversaries can gauge the other group's strength. They eventually all retreat while producing aggressive calls. In other encounters, some of which were observed, in other cases inferred, the casualties were counted. Eventually, the members of the splinter group were all hunted down and bitten to death.

Charles Darwin clearly did not accept the premises of the cult of primitivism that believed in the goodness of the "noble savage." But the account in *The Voyage of the* Beagle of his travels in Argentina with General Rosas's troops reveals that he did not expect "civilized" men to act any better. Juan Manuel de Rosas, the dictator who ruled Argentina from 1829 to 1852, organized a war of extermination, hunting down the indigenous population. Darwin met Rosas and rode with his troops when he explored the interior of Argentina, impressing them with his marksmanship as he collected game for his collections—and for the pot. Darwin never participated in any of their hunts of Indians.

POLITICS

Darwin was quite aware of the role of politics and connections in the human struggle for existence. He followed his grandfather Erasmus Darwin and his father, Robert Darwin, in being elected as a member of the Royal Society of London, England's foremost scientific society. Before the publication of *On the Origin of Species*, he served as secretary of the Geological Society of London and had corresponded and established connections with most of the leading scientists of his day. However, humans are not the only animals that play politics, cultivating acquaintances and forming alliances to achieve a goal.

Frans de Waal described his observations of chimpanzees living in a captive colony at the Arnhem Zoo in the Netherlands. Males can achieve alpha status by means of politics—gifts, favors, and fawning attention. Females also play chimpanzee politics. Chimpanzees have courts in which lesser chimpanzees groom their superiors, patting and smoothing their fur and removing insects, twigs, and other debris. The Gombe chimpanzees acted in similar fashion. Large powerful males tended to rely on violent brute force. Frodo, a large chimpanzee weighing over one hundred pounds, bit and hit his way to power; lesser chimpanzees groomed him. Frodo's court was similar to that of Ivan the Terrible, the sixteenth-century czar of Russia, though Frodo didn't execute his courtiers. Smaller males formed other coalitions in their quests to achieve alpha status, assiduously grooming other chimpanzees, both male and female. Wilkie, an eighty-pound weakling, groomed his way to alpha status. Chimpanzees, like humans, also can devise ingenious schemes to achieve power. Mike (the name Goodall gave to a small male in her Gombe study) intimidated other chimpanzees by charging while banging empty steel cans together. He was never observed actually attacking another male; the noise sufficed.

RECYCLING

William Paley's 1802 book on natural theology focused on the supposedly perfect "master plans" of God's creations. Paley introduced the image of God as a watchmaker, creating a device in which

> if the different parts had been differently shaped from what they are, or placed in any other manner or in any other order than that in

which they are placed, either no motion at all would have been carried out in the machine, or none which would have answered the use that is now served by it. . . . the watch must have had a maker. . . . who comprehended its construction and designed its use.[9]

Paley's book was required reading when Charles Darwin took his examination for the BA degree at Cambridge University. In his autobiography, Darwin stated that he was then convinced by the logic of Paley's arguments. But Paley's godlike watchmaker included defects in his creations: parts that frequently fail. Some, like the appendix in humans, don't do anything at all. And some features are self-destructive, such as irrational human aggressiveness. As Charles Darwin would later stress, no logic guides the opportunistic course of evolution.

That premise was evident when in 1838 Darwin concluded that natural selection was the primary agent in the transmutation of species. Throughout *On the Origin of Species* Darwin repeatedly emphasized the slow, gradual pace of natural selection, tied to the conditions of life in a particular time and place—a particular ecosystem. "I do believe that natural selection will always act very slowly, often over long intervals of time, and generally on only a few of the inhabitants of the same region at the same time. I further believe, that this very slow, intermittent action of natural selection accords perfectly well with what geology tells us of the rate and manner in which the inhabitants of this world have changed." Farther along in *Origin*, Darwin wrote, "she [natural selection] can never take a leap, but must advance by the shortest and slowest steps."[10] But Darwin was incorrect—natural selection can act quickly. In 1851, Charles and Emma Darwin's first child, Annie, died at age ten, probably from tuberculosis. Following the best medical advice, Darwin had taken her to a spa for a "water cure," but it had failed. Darwin and his contemporaries were unaware of the germ theory of disease. In Victorian England, tuberculosis was still a killer, but the Black Plague, those pandemics that had devastated England and Europe, were history. Selective sweeps—natural selection—had acted rapidly over a few generations and winnowed out virtually everyone who lacked natural immunity to the plague. Deadly pandemics had subsequently swept through indigenous American populations who lacked any resistance to endemic European diseases. Absent any knowledge of the role of bacterial and viral agents in the transmission of disease or the role of the immune system, Darwin could not have connected the absence of pandemics in nineteenth-century England with rapid natural selection. Had he been aware of the role of natural selection in sparing individuals whose immune systems could resist the bacterial and viral agents of the pandemics, he most likely would not have stressed natural selection's slow pace.

However, even if natural selection could act rapidly, it could only act so as to better adapt a species to life in a particular ecosystem. So how could his theory of evolution account for events such as the transition from aquatic to terrestrial life? Natural selection, rapid or slow, that created fish better adapted to aquatic life wouldn't enable them to live outside an aquatic ecosystem. Natural selection acting on variations that enabled a fish to swim faster, have sharper teeth, or exploit the lower depths of the sea would not help a fish to live out of water. Darwin's solution to this central problem was "recycling"—the reuse and modification of an existing organ.

In a few brief and often overlooked sentences deep in *On the Origin of Species*, Darwin pointed out that an organ "might be modified for some other and quite distinct purpose. The illustration of the swim bladder in fishes is a good one, because it shows us that an organ originally constructed for one purpose, namely flotation, may be converted into one for a wholly different purpose, namely respiration."[11]

Darwin drew on the research on lungfish that his close friend Richard Owen had published in 1839. African lungfish survive out of water because their swim bladders have been recycled. A few minor changes allowed lungfish to survive on land by reversing the flow of air stored in their swim bladders. Similar mudfish live in South America.

The original "purpose"—the selective advantage of a swim bladder—was to enable a fish to float at a given depth in the sea, a lake, or river while conserving energy. "Primitive" fish such as sharks have to swim constantly, expending energy. If a sharp stops swimming, it will sink. In most aquariums you can see sand sharks resting on the bottom of the tank. In shallow coastal regions, sharks can rest by lying on the seafloor. But in deep water a sinking shark can be crushed by water pressure, which increases with depth. In "advanced fish," natural selection crafted swim bladders, internal elastic sacks that can be filled with air to make a fish larger or smaller, so as to match the mass of displaced water at a depth and thereby effortlessly float. Goldfish hover in fishbowls thanks to their swim bladders.

In phylogenetically "advanced" fish such as goldfish, air extracted from the water by the fish's gills can be transferred into or back from elastic swim bladders, to make the goldfish's body larger or smaller in order to displace a mass of water equal to that of the fish. The net result is that a goldfish can hover, occasionally flapping its fins. It's the same mechanism that allows a blimp to hover in the air—the volume of air displaced by the blimp or balloon approximates the mass of the balloon or blimp. In a lungfish, an opening in the roof of the fish's mouth allows air to flow into its swim bladder when the fish is out of the water, transferring oxygen into the fish's bloodstream.

Recycling is a multistage process that involves both chance and natural selection. Ernst Mayr, one of creators of the twentieth-century "Synthetic Theory" of evolutionary biology that merged Darwin's insights with Mendelian genetics, coined the term "preadaptation" to refer to this process. Mayr pointed out his debt to Darwin. Stephen J. Gould and Elizabeth Vrba used the term "exaptation" to describe a similar evolutionary mechanism. In their widely cited paper "The Spandrels of San Marco and the Panglossian Paradigm," Gould and Richard Lewontin pointed out that the spandrels—arches that held up the cathedral of San Marco in Venice—served to provide surfaces that could be decorated. The further steps that followed from the cathedral having these serendipitous surfaces involved painting images or sculpting bas-reliefs. But as early as 1859, Charles Darwin had pointed out the role of recycling in the transitions that mark the course of evolution, though he didn't use that term.

OUR WACKY LUNGS

No one who understands how the lungs of humans or other mammals work would think that William Paley's "maker," that is, God, "comprehended its construction and designed its use." Our lungs are wacky devices that started as a recycled fish's swim bladder.

When we breathe in, taking air into our lungs during inspiration, there is no direct connection between the lungs and any of the muscles that inflate our two lung sacks. Instead, a sealed space, the plural cavity, encloses the elastic lung sacks. And we probably have two lungs because a fish would tend to spin on its long axis if it had but one swim bladder.

Muscles in the abdomen, chest, and the diaphragm are tensed so as to increase the volume of the body. This indirectly causes the lung sacks to expand and the air pressure in them to fall (as per the "Law of Physics" in the study Robert Boyle published in 1662).

Atmospheric air at a higher pressure then flows into the elastic lung sacks during inspiration. The elasticity of the two lung sacks then forces air out when the inspiratory muscles relax during expiration. But the sealed plural cavity must remain intact and not fill up with fluid. If that happens, death results.

You can perform the following experiment if you have a long, deep bathtub. Fill the tub with comfortably warm water. If the water is cold, the experiment probably won't make it to completion! Then immerse yourself, keeping only your head and one arm above water. Breathe in, and carefully mark the water level on the side of the tub. When you breathe out, the water level will be lower. You were larger when you inflated your lungs. If you were a fish, your lung-body morphology would make sense, allowing you to match your body mass to that of the water displaced at a given depth, letting you hover. Otherwise, it's a useless curiosity reflecting evolutionary recycling that occurred hundreds of millions of years ago.

TALKING AND SINGING

The fishy origins of the human respiratory system, which we share with all mammals, is also evident in the larynx—the "voice box." Victor Negus, the British surgeon who presented his research on the evolution of the larynx in books he published in 1929 and 1949, showed that the evolution of the human larynx started with a simple sphincter, a valve that a lungfish could close to seal off its airways when it was submerged in water. Profound changes occurred in the course of evolution to adapt the larynx to produce sounds for communication, and, as a result, you cannot seal your lungs from water by closing your larynx while diving underwater.

Negus compared the lungfish protolarynx to a purse string. Hardly any purses now close this way, but some small jewelry bags

do. A series of evolutionary adaptations yielded larynges that produced sound. Cartilage, muscles, and other soft tissue were added
to convert the even flow of air out of the lungs during expiration
into controlled periodic "puffs" of air that we perceive as the pitch of
a person or animal's voice. In frogs and other amphibians, the larynx has two flaps, each placed on the side of the airway leading to
the lungs. Frogs can tightly close their laryngeal flaps to prevent
water from entering when they dive underwater. But when they are
out of the water they can "phonate," producing a periodic sequence
of puffs of air by partially closing the two flaps and blowing air
through them. Further adaptations over the course of millions of
years added cartilage and muscles to form larynges like ours,
in which developed "vocal cords," complex structures of cartilage,
muscles, and other soft tissue whose tension and position can be
adjusted to phonate at different rates. The term "phonation" refers to
the acoustic source of energy, periodic puffs of air perceived as
sound, that the larynx can produce when air is blown through it.

The average fundamental frequency of phonation, the rate at
which the air puffs occur, is perceived as the pitch of your voice.
Many animals modulate the fundamental frequency of phonation
to produce calls that signal whether a predator is near, the availability of or desire for food, or to court potential mates. We humans
also signal emotion in this manner: variations in pitch can convey a
person's emotion and mood, and humans can also modulate the
fundamental frequency of phonation for linguistic ends. In the Chinese languages and in other tonal languages, words that have very
different meanings are differentiated by their pitch patterns.[12] All
known human languages also signal the end of a sentence by means
of distinctive pitch patterns, usually a sharp fall. Other linguistic
contrasts are transmitted by subtle, controlled acoustic "cues" that
involve coordinating the start and end of phonation with other
motor commands. The distinction between the initial consonants of

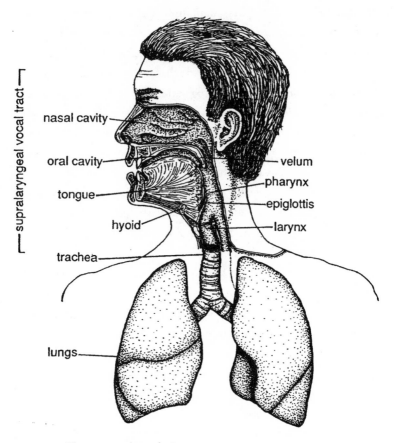

FIGURE 2.1 Human speech-producing anatomy

The lungs provide the airflow necessary to talk. The larynx, often called the "voice-box," is a complex structure. The "vocal cords" of the larynx are complex assemblages of cartilage, muscles, and other soft tissue. When we breathe, the vocal cords open wide. They narrow when we talk, and during *phonation* they rapidly open and close, releasing a stream of "puffs" of air. The rate at which the puffs of air occur, the fundamental frequency of *phonation* (F0), is perceived as the *pitch* of your voice. F0 variations can signal emotion, segment the speech signal into sentences, and, in *tone languages* such as Chinese, specify words. The supralaryngeal vocal tract (SVT), the airway above the larynx, filters the acoustic energy produced by the flow of air through the larynx to produce local energy maxima—*formant frequencies*. Formant frequencies play a central role in specifying consonants and vowels and achieving the high rate at which human speech transfers information.

the words *bat* and *pat*, for example, rests on whether phonation starts within twenty milliseconds after you open your lips.[13] The phonetic notation indicated by the bracketed [b] and [p] indicates these consonants. Similar differences in timing differentiate the initial [b]s and [p]s of other words.

VOCAL ACROBATICS

The maneuvers that have to be planned and executed when you talk or sing are not what one would expect had William Paley's watchmaker designed our lungs and larynx. Laryngeal physiology is such that the fundamental frequency of phonation will rise or fall when the alveolar air pressure (the air pressure in the lungs) rises or falls. This explains an aspect of respiratory control that appears to be a complex, innate, heritable adaptation for vocal communication. No one seems to have to learn to coordinate the necessary complex motor commands that allow people to phonate. They appear to be in place a few weeks after birth.

Breathing is necessary to sustain life, and when you take a breath, "inspiring" air, your lungs inflate. The inspiratory muscles of the abdomen and chest move outward, indirectly inflating the lungs. The diaphragm, the muscle that sits atop the stomach and digestive system, also moves downward during inspiration to expand the lungs indirectly. As the elastic lungs inflate, they act as springs, storing energy. During expiration, the elastic lungs are released, forcing air out of the lungs as they deflate. And like the air that flows out of an inflated rubber balloon when it is released, the air pressure in your lungs starts high and falls as the lungs deflate. The air pressure in a fully inflated balloon and in your lungs is at its maximum owing to the force generated by the stretched rubber or elastic lung sacks.

If you have ever blown up a balloon and released it (without tying a knot), you know that it will initially shoot around the room quickly but soon slow down as the air pressure generated by the force stored in the balloon's walls decreases. The balloon essentially is a spring that exerts maximum force when it's fully distended. As the balloon collapses, the force exerted on the air falls, and, as Isaac Newton's laws of motion predict, the balloon's velocity decreases.

The elastic recoil of the lungs provides most of the force that propels air out of them during expiration. We have comparatively few muscles that can directly expel air from our lungs. Emphysema, a disease in which the lungs lose their elasticity, therefore presents a serious medical issue—patients have difficulty exhaling. William Paley's clockmaker is a very "unintelligent" designer. It's always the case that during quiet respiration, the air pressure at the start of expiration is at its maximum and then gradually falls as the lungs deflate. But when anyone talks or sings, that is not the case—they unconsciously plan and execute a complex set of muscle commands keyed to the length of a musical phrase or sentence that they intend to sing or speak.

If you were to talk and did not oppose the elastic force of your distended elastic sacks when you attempted to utter a sentence, your pitch would be very high at the start of the sentence and then would fall throughout the sentence. Your vocal cords also might be kept pushed apart by the high air pressure at the start of the sentence, producing a raspy sound. These phenomena would mar talking or singing. Every song would start out at a high pitch or breathy shriek and then fall. Every sentence would have the same phonation pattern. Instead, whenever anyone intends to speak or sing, they "program" the muscles of the chest and abdomen to oppose gradually the force generated by the elastic recoil of the lungs. The abdominal and chest muscles that expand the lungs during inspiration are used for this purpose during vocal expiration to achieve a more or less even air pressure acting on the larynx. The result is that most sentences have a more or less steady pitch pattern that can be modulated upward for emphasis or to superimpose the tones that differentiate words in languages such as Northern Chinese, where, for example, the word *ma* has very four different meanings depending on its pitch. For most languages, the end of a declarative sentence is marked by the fundamental frequency falling abruptly—the orthographic period invented by scribes to indicate the end of a sentence. However, this again depends on maintaining a fairly stable and even alveolar air pressure throughout most of a sentence. To achieve this stable air-pressure "platform," your brain must unconsciously take account of the length of the musical line that you intend to sing or the length of the sentence that has been formed in your mind before you utter a sound.

During quiet breathing, the length of the inspiratory-expiratory cycle for a healthy person at rest is about two seconds. The durations of inspiration and expiration are about equal. However, when talking or singing, the duration of expiration is keyed to the length of the sentence or musical phrase. A longer sentence has a longer expiration. Before you produce a long sentence, you've generally decided to inflate your lungs to a greater degree. Without more air flowing through your lungs, you might pass out when declaiming at length. So your lungs have more air in them before you even start to utter a sound, yielding a higher elastic recoil–generated air pressure at the sentence's start. Less air is inspired before you start to utter a short sentence. Singing involves a similar pattern. What everyone unconsciously does is to program their "inspiratory" chest and abdominal muscles so as to oppose the elastic recoil–generated pressure keyed by the intended sentence or song's line length.

The diaphragm, though it is an inspiratory muscle that expands the lungs during quiet inspiration, is immobilized when anyone talks or sings, contrary to the traditional instructions to singers about controlling the diaphragm. The diaphragm has comparatively few "muscle spindles," sensors that can signal its position to the spinal cord and brain, which appears to limit its ability to execute rapid, fine adjustments during expiration.

In short, a longer sentence starts with more air in a person's lungs and a corresponding higher elastic recoil–generated pressure. But we program an opposing alveolar pressure stabilizing abdominal/chest muscle function that will yield a baseline pitch contour for any sentence. Similar maneuvers occur when anyone sings. These maneuvers, of which we have no conscious awareness, are innate human attributes that reflect the evolution of our recycled swim-bladder lungs. As is the case for other recycled organs, chance events started the process, and heritable adaptations were then passed on by natural selection. It's probable that apes and other mammals that produce long vocalizations, for example, wolves and whales, also have similar capabilities, but no one has yet looked into this matter.[14]

Victor Negus also pointed out that social animals that vocalize have larynges that trade off respiratory efficiency to enhance phonation. Horses have larynges that open wider than their windpipe (trachea). The larynges of cats and humans can open only halfway, which impedes airflow during exertion.

OUR RECYCLED TONGUE

Charles Darwin also took note of our recycled tongue and upper airway: "The strange fact that every particle of food and drink which we swallow has to pass over the orifice of the trachea, with some risk of falling into the lungs." Choking on food remains the fourth leading cause of accidental death in the United States.[15] This uniquely human problem arises because the human tongue has been recycled to enhance the process by which we perceive speech and communicate. The position in the neck of the human tongue and its shape has been modified so as to enable us to produce the vowels [i] and [u], the vowels of, for example, the words *see* and *too*.

The larynx, which rests atop the windpipe (the trachea), is anchored to the root of the tongue in all mammals. At birth, a human infant's tongue resembles a chimpanzee's and is mostly positioned in its mouth. Its shape is flat, as is the case for other mammals, and its primary purpose is to propel food entering the lips down the mouth. When drinking liquids, infants raise the larynx to form a sealed passage from the nose to the lungs, while the liquid flows around the larynx. Human infants, like virtually all mammals, essentially have two separate pathways running down their necks—one for breathing and one for drinking—because the air pathway from their nose is sealed off from their mouths. This explains why human infants can suckle milk and cats of all ages can lap up milk uninterrupted by breathing. Human infants also are "obligate" nose breathers. Their brains are programmed so that they can only breathe through their nose; they can't breathe through an open mouth. Pediatricians on call in hospitals usually carry a short length of plastic tubing. When a blueish-looking baby with a blocked nose arrives at the emergency room, the quick fix to get breathing going is to guide the tubing down the baby's mouth through the larynx.

A complex developmental process that spans the first eight years of life takes place in humans to arrive at adultlike morphology. Victor Negus began to describe the process, but the full details did not emerge not until 1994.[16] During this time, the neck becomes longer, and the skull changes its shape. The strange infants in Renaissance paintings often resemble shrunken adults because the painters didn't observe and paint infants. The neck gradually becomes longer as the tongue moves down into the throat, changing its shape to an

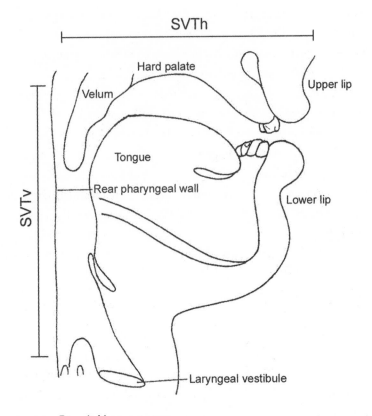

FIGURE 2.2 Recycled human tongue

At birth, the tongue is positioned almost entirely within the mouth, the *oral cavity*. The human tongue gradually changes its shape, so that by six to eight years of age half of the tongue is positioned in the mouth (SVTH) and half (SVTV) in the neck. As the root of the tongue moves down, it carries the larynx down. Darwin pointed out that the low position of the human larynx increases the risk of choking on food. The tongues of other mammals' are positioned in their mouths throughout life; their larynges, positioned closer to their mouths, are less likely to be blocked when they eat. The species-specific human tongue allows humans to produce "quantal" speech sounds that enhance the robustness of speech communication and its communicative speed.

almost circular posterior contour. The result, by the age of six to eight years, is a human airway located between a person's lips and larynx, the human "supralaryngeal vocal tract" (SVT). Half of your tongue is positioned in the mouth—the "horizontal" (SVTH) section. The other half of your tongue is positioned vertically (SVTV) down the throat, in the pharynx. The oral and pharyngeal sections, SVTH and SVTV, meet at an approximate right angle, owing to the tongue's almost circular posterior profile.

Victor Negus thought that the unique shape of the human tongue somehow facilitated speech communication, an advantage compensating for the increased risk of choking. Negus couldn't test his theory, but it has been confirmed by independent computer-modeling studies that calculate the range of formant frequencies that could be produced by human and nonhuman SVTs. In 1969, my colleagues and I used a computer-modeling technique to determine the vowels that a rhesus monkey's SVT could produce.[17] The range of tongue shapes was estimated by taking into account constraints on tongue deformation. The monkey's tongue was positioned to best approximate the SVT configurations used by adult human speakers to produce the "point" vowels [i], [u], and [a]. These vowels occur and delimit the range of vowels of all human languages.

The human tongue's oral and pharyngeal proportions and shape explain why only humans older than six to eight years can produce the vowels [i], [u], and [a]—and why these vowels contribute to the robustness of human vocal communication. The extrinsic muscles of the tongue, anchored in bone, can move the tongue about as a whole to create abrupt midpoint ten-to-one discontinuities in the SVT's cross-sectional area. In 1972, Kenneth Stevens's parallel research explained why the unique human tongue contributed to the robustness of human vocal communication. Stevens showed that only the species-specific human SVT can produce the ten-to-one midpoint-area function discontinuities, abrupt changes in the cross-sectional area of the vocal tract, that are absolutely necessary to produce the vowels [i], [u], and [a], which he termed "quantal." Stevens used computer-modeling techniques that predict the acoustic effects of sound passing through oddly shaped tubes. He also used wooden tubes whose shapes could be changed so as to change the position of the 10:1 changes in the SVT cross-sectional area. The quantal vowels are perceptually salient owing to their having spectral peaks, akin to saturated colors. Their acoustic signatures also do not shift when a speaker makes slight errors, shifting tongue position about one centimeter around the midpoint. Speakers thus can be sloppy yet produce the "same" vowel. Terrance Nearey subsequently showed that the vowel [i] is an optimal signal for determining the length of a speaker's vocal tract—a necessary step in the complex process of recovering the linguistic content from the acoustic signals that convey speech. Whereas the identical acoustic signal can represent two different vowels for speakers who have different SVT lengths, no such overlap occurs for instances of [i].

The basis for these effects rests in the physiology of speech production, which also makes human speech "special." It became evident in the 1960s that human speech makes complex human language possible. The research group directed by Alvin Liberman at Haskins Laboratories, then located in New York City, was attempting to make a text-to-speech system that would read printed text to blind persons. The Haskins group had to depart from received linguistic theory, which held that phonemes (similar to the letters of the alphabet) can be freely reordered to form different words, as though they were movable type. Their seminal 1967 paper, "The Perception of the Speech Code," showed that phonemes did not exist in the stream of speech. The then-new technology of tape recording allowed them to attempt to segment the acoustic speech signal into phonemes. If the speech signal actually consisted of a string of phonemes, then a tape recording of someone uttering the word *too* should have had a segment of tape that contained the phoneme [t], followed by a segment of tape that contained the vowel [u]. (The symbol [u] represents the word *too*'s vowel in standard phonetic notation.) However, when the ostensible [t] segment was excised from the word *too* and was placed before a tape segment excised from the word *tea* that contained its [i] vowel, the result was incomprehensible. Similar results held for all "stop" consonants (in English, the phonemes [b], [p], [d], [t], [k]) and other vowels.

It became evident that words were produced as complete "encoded" entities. Each articulatory gesture fused into the adjacent hypothetical segments. For example, when uttering the word *too*, all speakers protrude and narrow their lips at the very start of the word, anticipating the following vowel [u]. In contrast, when uttering the word *tea* all speakers retract their lips at the very beginning of the word, anticipating the following vowel [i]. Human speakers aim at producing complete words. Discrete segmental phonemes—"beads on a string" that form words equivalent to the letters of the alphabet—do not exist in the acoustic signal. Phonemes are abstract entities, perhaps instantiated in the brain as control patterns in the parts of the cortex implicated in motor control.

In short, we can talk rapidly and get all the words of a sentence into the short-term memory of whomever we are talking to via this complex, nonintuitive process. We do not directly decode the flow of speech into phonemes. We instead recognize words.

All speech-recognition systems work on this basis. Attempts to develop computer systems that would identify phonemes failed decades ago. It would be impossible to understand what anyone is saying if we actually had to recover phonemes from an acoustic speech signal. At normal rates of speech, about twenty to thirty hypothetical phonemes would be transmitted per second. This rate is faster than the fusion frequency of our auditory system, at which sounds merge into an incomprehensible buzz. If we actually were transmitting individual phonemes, we would be limited to perhaps seven per second. We would forget the beginning of a sentence before we came to its end. Ignoring these facts, linguists almost universally hold to the theory that permutable phonemes, beads on a string, are one of the defining features of "articulated" human language.

However, engineers, who have to make things that work, such as smartphones, recognize what you're saying, matching the incoming acoustic signal usually against word-level templates that take into account differences in vocal-tract length and dialect. Alphabetic notation is only one possible writing system. The Chinese languages have for thousands of years transcribed words as entities. You, the reader, may think that discrete phonemes really exist because alphabetic systems seem "natural" to you. That wouldn't necessarily be the case if you were Chinese.

JOHANNES MULLER AND THE SOURCE-FILTER THEORY OF SPEECH

The basic mechanisms of speech production were described in 1824 by Johannes Muller, one of the founders of physiology and psychology. A pipe organ works in essentially the same manner to produce musical notes. The world's oldest existing playable pipe organ, in Sion, Switzerland, was built in 1435; earlier organs probably existed. Some knowledge of speech physiology, therefore, was present in medieval Europe.

A bellows provided the necessary airflow, which generated an acoustic source that had energy over a broad range of frequencies. When a person presses a key on the organ's keyboard, a valve opens a pipe of a particular length and shape—it is either open at both ends or closed at one end. The pipe acts as an acoustic filter, allowing maximum energy to pass through it at frequencies determined by the length and shape of the pipe, producing a musical note. The colors that you see through a pair of sunglasses are determined by a similar process. Sunlight has energy distributed across the visible frequency range of electromagnetic energy. The tint of the sunglass lenses reduces energy at specific frequencies, and the range of frequencies that pass through the sunglasses determines whether everything that you see is blueish, reddish, greyed down, etc.

The frequencies at which maximum acoustic energy can pass through the supralaryngeal vocal tract play a major role in specifying the vowels and consonants of speech. The notes that a pipe organ produces likewise are determined by the frequencies at which maximum energy can pass through a given organ pipe. These key frequencies are termed *formant frequencies*, and they must be recovered in the complex process by which we understand what's being said. Perceiving the message conveyed when someone is talking is not as simple as following the fundamental frequency patterns of birdsong or the pitch patterns generated by the larynx. It first is necessary to estimate the length of the SVT of the person who is speaking because the absolute values of the formant frequencies for any phoneme depend on the length of a speaker's SVT. That's because the SVTs of different speakers vary in length. Peterson and Barney, in one of the earliest studies directed at producing a machine that would recognize speech, found that the average formant frequencies of the vowel [i], for example, are 270, 2,300, and 3,000 Hz, while the formant frequencies of the vowel [u] were 300, 870, and 2,240 Hz, for a large sample of adult males.[18] The formant frequencies of both [i] and [u] would be 1.5 times higher for a child whose SVT length was 11.3 cm long than for an adult whose SVT was 17 cm long. However, the formant-frequency patterns of the child and adult male would be perceived as examples of [i] and [u] to anyone listening, owing to perceptual "normalization"—human listeners unconsciously estimate the length of the SVT of the speaker to whom they are listening.

Listeners can estimate SVT length after hearing a short stretch of speech or after hearing a known phrase or word, such as a person saying *hello*. Terrance Nearey in a 1978 study showed that the vowel [i] (of the word *see*) was an optimal signal for immediate SVT normalization. Other sounds will do, but [i] works best. In the 1950 study by Peterson and Barney aimed at machine recognition of speech, two errors in 10,000 trials occurred when panels of listeners had to identify words that had the form consonant-vowel-consonant if the central vowel was an [i] when they had to adjust instantly to a different speaker's voice. Six errors occurred for the vowel [u]. Other vowels resulted in hundreds of errors.

In short, perceiving spoken words entails a listener unconsciously estimating the length of the SVT of whoever's talking. This mental operation has its roots in animal communication. Many species estimate the size of conspecifics by taking account of the absolute value of the formant frequencies of their cries. Vocal-tract length in these species correlates with body size. Hence longer SVTs produce lower formant frequencies, providing an acoustic marker of body size. This may account for dogs being able to recognize words. Formal tests confirm what anyone who has had a dog knows: dogs understand what you are saying—when it concerns them. (We'll return to what dogs understand later.)[19]

Our recycled tongue, which allows us to produce quantal vowels, plays a key role in this process.

REMEMBRANCES, INSTINCT, AND RECYCLING

Marcel Proust, in his seven-volume novel *Remembrance of Things Past*, described scenes and images of past events in detail. Current research shows that these visual "remembrances" entail using areas of the visual cortex that until recently were thought to be involved only in perception—seeing an object or scene.

When you look around at the world, the shapes and colors of both stationary and moving objects, people, animals, the sky, and so on appear as a single image. You might suppose that your brain has an "organ," a "Faculty of Vision," in which the image projected on the retina of the eye is somehow captured and stored. The obvious analogy would be the image captured on the sensor of a digital camera, which can be transferred whole to a computer screen or printed out. That, however, is not the case. It has become evident that neural circuits, pathways in the brain that link activity in a series of neural structures, sequentially act on the signal transmitted from the retinas of an animal's eyes. Some of these structures are connected in an "upper," "dorsal" pathway in the cortex and seem to be involved in following motion and determining "where" things are. A lower, "ventral" pathway seems to be involved in identifying and storing "what" you are viewing. Primary visual cortex, termed area V1, receives the signal from your eyes. V1, oddly, is not located close to your eyes but at the very back of your brain.

In the early 1960s, David Hubel and Torsten Wiesel used micro-electrodes—exceedingly thin devices that could record electrical activity in the brain—to study the responses of individual cells in V1 in cats. Unlike many of the cells in the retina, which respond to the luminosity of spots, whether they are light or dark, they found that cells in the visual cortex were highly selective for edges or lines at specific orientations.

The pattern of electrical activity that they monitored did not suggest that V1 was recording anything that resembled the complete image, as would a camera's film or a digital sensor. A horizontal line, for example, would produce the same response no matter where the cat saw it. The output of area V1 went into cortical area V2, then V3, and so on, areas that later research showed responded to colors and shapes; this was all somehow integrated in the brain to form the images that cats, people, and other animals see. The receptors in V1 had to be activated within a "sensitive" period for normal vision to occur. Kittens deprived of vision during a critical period also never fully developed normal activity in V1 and had profound difficulty in interpreting the visual world. Hubel and Wiesel received the Nobel Prize in physiology and medicine in 1981.

Digital computers were just coming into general use at the time, and it was assumed that the functional architecture of the brain was more or less similar to a computer system. In most digital computers, discrete modules are devoted to specific tasks. The electrical circuits and components that process incoming images, for example, are devoted to that task, and a different module—a different circuit linking different electrical devices—displays those images on the computer's screen. It has since become apparent that the functional architecture of biological brains differs from that of conventional digital computers. "Organs" of the brain that are involved in perceiving images and sequences of images also form part of the brain's visual memory, starting with primary visual cortex, V1. And these studies address general questions that Darwin opened, questions about the neural bases of instincts and the mechanisms by which living organisms learn about the world.

HOW A MOUSE VIEWS THE WORLD

Although mice and other rodents have very small brains compared to humans, elephants, and whales, they are the most ubiquitous and successful mammals on earth. Their survival depends on learning to recognize the features of the visual world so as to negotiate it to find food and avoid predators. In both of these critical activities, it is useful to be able to recognize familiar objects and events so as to respond to something novel. Unlikely though it seems, how mice view the world bears on the question of whether humans are born with an innate knowledge of physics, language, or morality.

The evidence cited for humans being born with an innate knowledge of physics hinges on what six-to-eleven-month-old infants do when they view videos. If they watch the video multiple times or if it shows something very familiar, the infants become "habituated," that is, bored. Experiments that monitor the eye movements of six-month-old infants watching videos showing a ball dropping versus a ball suddenly rising have been interpreted as evidence for "innate physics." The infants rapidly become bored watching the familiar sight of a ball falling down, whereas they snap to attention when a ball rises upward, defying gravity. Six-month-old infants also pay closer attention to computer-generated videos that show people walking through walls than to videos of someone stopping when they reach the wall. The responses of these infants are often interpreted as evidence for humans having an instinctive, innate, "core" knowledge of physics.[20]

According to "nativist" theoreticians, our brains at birth have a genetically transmitted core knowledge of physics and of the strength of materials. Newton should not have wondered why objects fall down. If innate physics was already coded in his brain, it wasn't worthy of notice or thought. But cannot infants simply repeatedly observe and hence learn that balls fall downward and that people can't walk through walls? Studies on mice suggest that is the case—and that a "recycling" of primary visual cortex, area V1, is involved.

Mark Bear and his colleagues, and students at MIT's "Bear Lab,"[21] have shown that primary visual cortex V1 of mice has been recycled to take on an additional function. V1 serves as the input to the mouse brain's visual cortex, but it also is involved in learning to recognize visual patterns as well as temporal sequences of images—examples of syntax akin to that of human language. The Bear research group monitored electrical activity in V1 using microelectrodes while simultaneously monitoring the fidgeting of the mice's paws. They had previously found that mice spontaneously fidget in response to visual stimuli. Different electrical and fidgeting response patterns occurred when the mice viewed images that were familiar or novel. When mice were free to roam about their cages, they spent more time responding to new visual patterns. When the mice were presented with various four-pattern sequences, they also learned to recognize familiar sequences.

By using hormonal "inhibitors" that blocked V1 activity, Bear's research group was able to show that experience-dependent activity in primary visual cortex V1 was responsible for the mice learning to recognize these stimuli. By the age of six months, human infants have seen many objects falling down and perhaps a few helium balloons at parties going up. And they haven't seen anyone walking through a wall.

Many other areas of the cortex are involved in storing and recovering images, words, and events. However, as Mark Bear's research on area V1 shows, neural memory is not a "module" similar in any meaningful way to the hard drive in your computer. The functional architecture of a human, mouse, or any mammalian brain—and probably similar systems in animals like birds that don't a possess mammalian cortex—does not involve a CPU that rapidly performs computations and a hard drive that stores, that "remembers," information. It's unclear how and what various parts of the cortex are doing in humans and other animals, but it is clear that brains work as dynamic systems.

It's clear that having a bigger brain, including a vastly larger cortex, plays a significant role in making animals, including humans, smarter. In 1699, Edward Tyson, in the first anatomical study of an ape, demonstrated many of the anatomical similarities between an orangutan and humans, but it was apparent that the ape's brain was much smaller than that of any normal human adult.

In 1868, shortly after the publication of *On the Origin of Species*, Louis Lartet noted an evolutionary "cognitive arms race" between wolves and their prey. The skulls of wolves and sheep who died long ago had been preserved, buried in layers of sedimentary earth deposited by the Seine river. The depth of each layer provided a rough measure of time. Skulls found in deeper levels were those of animals that had lived at earlier times. Wolves are cleverer than sheep, and, not surprisingly, the wolves living at any given time had bigger brains than their meals. However, Lartet discovered that over time the brains of both sheep and wolves had evolved, becoming larger. Surprisingly, the brains of sheep closer in time to 1868 were bigger than the brains of earlier wolves. The wolf-sheep brain-size race makes sense if having a bigger brain increases cognitive ability. In all periods, some of the smarter sheep had outwitted wolves and survived, and their lambs had retained their sires' and dams' bigger brains. But wolves maintained their edge. Since Darwin's time, many studies have shown that supporting a larger brain requires having a body that generates more energy—dozens of books, including some of my own, have pointed this out. Again, as Darwin stressed, a synergy exists between how a species interacts with the environment and biological evolution. As John Shea points out, hominins millions of years ago began to produce stonecutting tools, which increase the efficiency of eating meat or any food, so as to generate more energy, which in turn would support larger brains.[22] Fire and cooking likewise provide more nutrients. Chapter 4 will follow up on the functional architecture and evolution of the human brain.

THE CORTEX

Over the course of millions of years, hominin brains have increased in size, even taking into account increases in overall body weight and height. Brains of the fossil hominins discussed in the next chapter show that increases in brain size roughly track the complexity of the stone tools that they produced. The human cortex is almost three times the size of an ape's. The posterior temporal region is disproportionately large and, as many studies suggest, may signify an increase in storage space. As is the case for digital computers, increased storage space can result in a biological "computer" that can execute qualitatively different tasks. The first digital computers that I used had tiny memory-storage capacities (one is now in a computer museum) and were incapable of performing the tasks now easily carried out by the "apps" on my smartphone—voice-activated commands were out of the question and the first attempts at AI— artificial intelligence—were a joke. Aside from faster processors, the primary difference between your laptop or a contemporary supercomputer and the computers of the 1960s and 1970s is memory size. The threefold increase in the size of the human cortex thus is significant. But as will be apparent in chapter 4, though no one can state with certainty what the different parts of the cortex do, the aspects of behavior that we can observe appear to involve neural circuits that link activity in the cortex with subcortical structures that we share with mammals, lizards, amphibians, and other

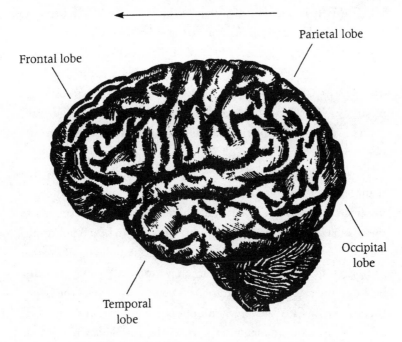

Frontal lobe

Parietal lobe

Occipital
lobe

Temporal
lobe

FIGURE 2.3 Simplified map of the human cortex

The cortex is divided into two hemispheres. The surface of the left hemisphere is shown here. Folds further divide the cortex into an "anterior" frontal region and posterior regions. There is no apparent "logical" design to the brain. Primary visual cortex (area V1), which receives information from frontal cortex, is located at its rear.

species. Virtually all parts of the human brain appear to have been recycled and modified to work more efficiently or faster.

Noninvasive neuroimaging studies consistently show that one area in the posterior left temporal cortex, the fusiform gyrus (FFA), responds when people look at faces, hence the acronym FFA—fusiform face area. In humans, the FFA is roughly in the lower midpoint of the brain. Noninvasive imaging techniques such as functional magnetic-resonance imaging (fMRI) have made it possible to monitor the human brain without inserting electrodes into it. MRIs image the soft tissue of the brain, and fMRI specifically tracks the level of oxygen in the blood—when neurons are active they "burn" glucose and deplete local oxygen levels. Hence depleted oxygen levels indirectly signal increased neuronal activity. The FFA was thought to be an innate human face–recognition organ by Nancy Kanwisher at MIT, who still holds to that view. Individuals who had suffered damage to the FFA had difficulty recognizing faces.[23]

However, a Scottish sheep herder who had suffered a stroke that damaged his FFA could no longer recognize the individual sheep in his flock. It became evident that the fusiform gyrus responds to images that are of interest to "experts." It seems to be a brain mechanism that responds in a decisive "categorical" manner to all manner of images that people want to become expert at assessing—cars, birds, or any category of weird objects you can imagine. The hypothetical innate FFA "face-recognition" organ located on the rear left side of your brain (the temporal cortex) also responds when people who are car or bird experts look at pictures of cars or birds.

One of our close friends is an ardent birdwatcher who can identify a bird's species instantly on hearing its call and can tell exactly where it is—while my wife and I haplessly peer here and there through our binoculars. When our friend was tested in Mike Tarr's laboratory at Brown University in an fMRI experiment, the cortical area adjacent to the FFA was activated when she listened to bird calls. That wasn't surprising, since the temporal cortex is a "multisensory" region of the brain, responding in perceptual studies to visual and auditory stimuli. As the brain bases of vision, hearing, and touch are being explored, it's becoming evident that a complex web of circuits that link activity in many parts of the brain is involved. Simplistic neophrenological "organs" or "modules" are not the answer.

INSTINCT AND LOVE

Darwin used the term "instinct" to refer to a behavior that he thought was heritable, hence innate. A feature attraction of Galapagos tours is the tortoise "finch-pose." A symbiotic relationship exists in the Galapagos Islands between tortoises and finches. The finches eat parasites on the tortoises' skins, thus ridding the tortoises of parasites. A Galapagos tortoise signals when it wants a finch to clean its skin by rising up on its short legs as high as possible while extending its neck upward. A finch then lands on its body and systematically pecks off and eats parasites on the tortoise's neck, legs, and exposed skin. The tortoise stays "finch-posed" until the finch completes its meal. The finch-pose seems to be controlled by an innate mechanism that evolved in the Galapagos to favor the survival of tortoises who were cleansed of parasites. The finch-pose probably is innate and shaped by natural selection. But it has been "recycled" to serve a very different purpose in a different ecosystem.

The San Diego Zoo does not have any Galapagos finches, but it does have Galapagos tortoises and plenty of people. The recycled Galapagos finch-pose is a featured zoo attraction. The Galapagos tortoises instead assume the finch-pose to entice their keepers to scratch their necks. They probably would finch-pose to everyone visiting the zoo. If someone wished to explore the neural bases for the tortoise finch-pose, it might result in a journal paper, but it hardly seems to merit poking electrodes into tortoise brains. The finch-pose seems be an innate mechanism that has been "recycled" to serve another end—scratch my neck—and it also might be viewed as a sort of mark of affection.

Mammals clearly show affection—"love"—which is an innate instinct. Mammalian affectionate behavior most likely derives from their having to care for helpless infants. In mice, the anterior cingulate gyrus, a structure of the "old" paleocortex that humans retain, controls mother-infant care-giving interactions. Paul Maclean in 1982 showed that mice don't pay attention to their infants when the cingulate or the neural circuits that connect it to other structures of the mouse brain are disrupted. Cat fanciers are certain that their cats love them, curling up and purring as they are gently scratched. Cats can't talk, so we can't ask them if they love us. The cat's response pattern may also reflect a built-in, genetically transmitted innate motor response that evolved in feral cats cleaning each other to eliminate parasites. However, it too has been recycled and serves to show a cat's affection to a human. In a sense, a cat may be able to love people. Dogs clearly bond with people, and I won't waste your time repeating real-life accounts of how dogs display affection. Horses often show affection to other horses, nuzzling and licking them, and they also display affection to people.

I became a horse doctor in 1998, on a journey in the foothills of the Himalayas in Ladakh, India. The highest mountains were about six thousand meters. My wife, Marcia, Lama Ngawang Jorden, a Buddhist monk, and I were traveling on foot because there were no roads leading to the monasteries we wished to reach. The Tibetan world extends beyond Nepal well into India. Our intent was a preliminary photographic survey of the wall paintings in these isolated Buddhist monasteries. Marcia had met Lama Jorden in Cambridge, Massachusetts, where he was completing his Ph.D. in religious studies at Harvard University.

Lama Jorden holds the rank of "Geshe" in Tibetan Buddhist scholarship, the equivalent of a Ph.D. He was the key that gained entry to every monastery and permission for me to photograph the wall paintings. We also became accustomed to Tibetan tea—a mixture of butter, salt, tsampa (finely ground parched barley), and boiled brick tea—served with cookies to us as honored guests. It is quite palatable—if you think of it as soup. In monasteries that were accustomed to British visitors, Darjeeling tea was served.

We were traveling with a guide, two cooks, and three small horses. The sure-footed horses, descendants of the ones that the Mongol hordes once rode, carried our gear. The trails were steep and rugged and often were cut into the sides of slopes high above swift-running rivers. The soil was sandy. At a few points the trail vanished, and there were short blank sections that you had to traverse carefully. Moti was the name of the horse who always walked with us, carrying our packs and medical and photo gear. The other horses would walk ahead with the two cooks, who set up camp for the night, boiled water, and prepared dinner.

Everything went along routinely until about a week into the trip, when at a one-meter-long break in the trail Moti panicked. Instead of following us over the gap, as he usually did, Moti instead tried to climb the slope to reach what he must have thought was a path. Instead, he slipped, tumbling head over hooves, rolling down the steep slope. We watched helplessly, frozen with fear, because below the trail was a two-hundred-meter sheer drop down to the Zanskar river. Fortunately, Moti landed twenty meters down the slope in a hollow. Our guide rescued him, but dirt and stone fragments were embedded in his chest, legs, and back. And I became a horse doctor.

Lama Jorden, Marcia, and our guide firmly gripped Moti while I started to remove the stones with tweezers, but Moti stood quietly while I washed and dressed his wounds. It must have been painful, but he seemed to realize that we were helping him. The problem then became how to keep the dressings in place. For the next two weeks Moti was the celebrated silver horse, banded with swaths of silver duct tape. Finally, the day came when Moti was pronounced cured, and the duct tape was removed. I was sitting eating my lunch, a chapatti sandwich, and I jumped up because I suddenly felt as though my right ear was being sandpapered by something very wet. Moti was licking me. He was recycling whatever faculty of horse nuzzling might exist to signal affection.

IN SHORT

The films on evolution that start by showing two beasts pouncing on each other do not reflect Charles Darwin's meaning when referred to "the struggle for existence." He instead stressed the mutual interdependence of species and their interaction with the ecosystems that they shared.

Intelligent design, the claim that the path of evolution is directed by God, derives from Robert Chambers's 1844 book *The Vestiges of the Natural History of Creation*. Chambers also promoted phrenology, which claimed that "Faculties" located in bumps on a person's skull conferred language, morality, arithmetic, etc. Both premises are still promoted today.

Darwin introduced the concept of "recycling"—the redirection and modification of an existing organ to achieve a new end. The wacky design of our respiratory system reflects the starting point— the recycled swim bladders of fish. Being able to talk—a quintessential human capability—involves having a recycled tongue. Human speech achieves a high data-transmission rate, one that exceeds the frequency at which other sounds merge into an indistinct buzz, by transmitting the words as entities rather than as strings of the "phonemes" hypothesized by traditional linguistic theories—elements that roughly approximate the letters of the alphabet. Phonemes don't exist in the speech stream. A complex neural process that involves estimating the length of a speaker's airway above a speaker's larynx is necessary to recover the meaning of a sentence. The recovery process is facilitated by our recycled tongues, which can produce speech sounds providing "honest" signals that help listeners estimate the speaker's airway length. The tradeoff, as Charles Darwin noted in 1859, is that our peculiar recycled tongues increase the likelihood of choking on food.

Structures of the cortex that are involved in the initial stages of vision in both mice and men have been recycled to serve to code memories of what we and mice see. In the chapters that follow, it will become apparent that the larger human cortex and recycled neural structures such as the basal ganglia play critical roles in the neural circuits that confer the motor, cognitive, and other capabilities that distinguish humans from other living species.

3

GRANDFATHER ERASMUS

Charles Darwin never knew his grandfather Erasmus Darwin, who died seven years before Charles was born. Erasmus was a polymath: a botanist, naturalist, inventor, and philosopher as well as a successful doctor and businessman. He was politically progressive and advocated measures aimed at improving public health.

Erasmus had six children with his first wife; three survived, including Charles Darwin's father. Erasmus also had the free and easy views on sex of eighteenth-century England. In the interval between the death of his first wife and his second marriage at age fifty-five, he fathered two children with his children's eighteen-year-old governess. He had another seven children with his second wife. Erasmus was the author of *Zoonomia; or, The Laws of Organic Life*, published in 1794 and 1796.

Erasmus Darwin's contribution to the theory that his grandson Charles presented in 1859 will be discussed in the examples below. We will also look at issues that Erasmus and Charles were not able to address fully, given the absence, at that time, of evidence in the fossil record for hominin evolution and comparative studies of the behavior of apes, our closest living relatives.

* * *

Zoonomia is a rambling book that catalogued diseases, their causes, and their treatment. It presented Erasmus's observations of embryological development, which played a central role in his theory of evolution as well as in Charles's later theory. Erasmus departed from the religious orthodoxy that asserted that God had created all forms of life in the Garden of Eden four thousand years ago. He thought instead that the earth was millions of years old. In Erasmus's theory of evolution, a species could be "transmuted" by inheriting acquired characteristics. Erasmus proposed that species evolve through interactions with the environment: "all animals undergo perpetual transformations; which are in part produced by their own exertions in consequence of their desires and aversions, of their pleasures and their pains, or of irritations, or of associations; and many of these acquired forms or propensities are transmitted to their posterity."[1]

Erasmus proposed that all animals evolved through a single line of ancestry, "a filament" of species increasing in complexity from a single common ancestor. Plants and cold-blooded animals also derived from this filament. Erasmus's basket of supporting data included observations of the comparative anatomy of animals, the structure of plants, embryology, and the known fossil record. However, *Zoonomia* was also packed with speculation. Though Erasmus's own life should have suggested that success is not always to the strongest, he proposed that "the strongest and most active animal should propagate the species which should thus be improved." In his posthumously published book in verse, *The Temple of Nature*, Erasmus attempted to reconcile his views on evolution with "Natural Law," the deist position that posited that whatever changes may occur follow a divine plan.

Erasmus Darwin has been forgotten, although he was for a time known as the "English Lamarck." Jean-Baptiste Lamarck, in his *Philosophie zoologique*, adopted the same position as Erasmus and again proposed that an animal could pass on to its offspring acquired characteristics. The "Just-So" story of how giraffes got long necks perhaps is the most familiar "explanation" of the Erasmus Darwin–Lamarck evolutionary mechanism. As giraffes stretched their necks to feed on leaves on higher branches their necks got longer, and their offspring inherited their longer necks. Curiously, you can find Creationist websites that focus on the giraffe just-so story, arguing that God instead must have given giraffes their long necks.

In 1859, Charles Darwin cited Lamarck and endorsed the views that acquired characteristics could be inherited, views derived from *Zoonomia*, as well as Erasmus's ideas concerning embryology and common ancestry, ideas that also play a central role in *On the Origin of Species*. However, Charles Darwin did not cite *Zoonomia*—even though Charles's original title for the first draft of *On the Origin of Species* was *Zoonomia*.

In Charles Darwin's letters and in the autobiography that he wrote for his children and grandchildren, he claimed that, though he had read *Zoonomia*, it had produced "no effect on me." In light of Victorian attitudes toward extramarital sex, Charles Darwin perhaps wished to bury all memory of Erasmus. Or he might have wished to distance his book from the speculative tone of *Zoonomia* and the deist position of *The Temple of Nature*, in which evolution was guided by a master plan. Charles instead gathered his own data and performed experiments to support his claims. He also was able to cite the detailed embryological studies that had been published since Erasmus's time. However, as Janet Browne has pointed out, Darwin in his autobiography "conveyed a decidedly self-congratulatory element." His theory was crafted entirely by him and owed nothing to grandfather Erasmus.[2] But that clearly was not the case.

COMMON ANCESTRY

When I was a child, Coney Island was the playground of New York City. Coney Island's wide sandy beaches, boardwalk, roller coasters, parachute jump, amusement parks, and hot-dog stands probably are what people of a certain age recall. But the Coney Island that I knew had unpaved side streets on which my playmates and I would build small fires and roast potatoes, which we thought were a treat. We also staged elaborate games that involved trading marbles, and we fearlessly roller-skated on the paved streets—cars were very few. We were almost all the first-generation children of either Italians who had emigrated from Sicily or Eastern European Jews.

On lazy summer afternoons, the Sicilian grandfathers gathered beneath trellises bearing wine grapes as they watched their fancy pigeons fly above the rooftops. They had built large pigeon coops, and the pigeons would be released in the late afternoon. The birds would ascend, circling, tumbling, and swooping. The game then played was to see to which coop each pigeon would return. The birds were more or less sorted out over tumblers of homemade wine and little slices of cake. We children only had cake.

The pigeons were not the dull, drab street pigeons that now inhabit the streets of New York. They were "true-breeding" fancy pigeons: fantails, pouters, Jacobins, Giant Runts, homing pigeons, tumblers—the breeds kept by serious pigeon fanciers. The pouters had huge crops that they continually inflated. The fantails had profuse tail feathers. Tumblers would fly up in a close group and when descending fall heads over tails.

On the Breeds of the Domestic Pigeon

Charles Darwin did not hesitate to conduct "fool's experiments" that explored unlikely possibilities. In his biography of his father, Francis Darwin recounted Charles's

> love of experiment.... He was willing to test what would seem to most people not at all worth testing. These rather wild trials he called "fool's experiments," and enjoyed extremely. As an example I may mention that finding the cotyledons of Biophytum to be highly sensitive to vibrations of the table, he fancied that they might perceive the vibrations of sound, and therefore made me play my bassoon close to a plant.[3]

The seed leaves, cotyledons, of the "little tree plant" *Biophytum*, native to Southeast Asia, did not move when Francis played his bassoon. In that instance, Darwin drew a blank.

Regarding pigeons, Darwin had greater success. As Ernst Mayr, one of the central figures of twentieth-century evolutionary biology, has pointed out, Darwin was one of the first to practice the method of modern science.[4] Darwin started by forming a preliminary theory on the basis of initial observations. He then tested the theory against the findings of experiments aimed at refuting it, and when necessary he revised his theory. Erasmus Darwin had proposed a common ancestor for all species, and Charles tested a special case of Erasmus's theory—the ancestry of the true-breeding fancy pigeons that flew above my home in Coney Island.

Shortly after he returned home from his voyage on the *Beagle*, Darwin had plunged into the world of pigeon breeding, which then was a gentleman's fancy.

> Believing that it is always best to study some special group, I have after deliberation taken up domestic pigeons. I have kept every breed which I could purchase or obtain, and have been most kindly favoured

with skins from several quarters of the world. . . . Many treatises in different languages have been published on pigeons, and some of them are very important, as being of considerable antiquity. I have associated with several eminent pigeon fanciers, and have been permitted to join two of the London Pigeon Clubs. The diversity of the breeds is astonishing.[5]

Darwin took detailed notes on the different breeds' plumage, skeletal structures, eyelids, toes, skin, flight paths, calls, and habits. The pigeon fanciers that Darwin consulted believed that their true-breeding pigeon breeds had each descended from a different species. Their hypothesis was reasonable, since, as Darwin noted, the different pigeon breeds had attributes such that

at least a score of pigeons might be chosen, which if shown to an ornithologist and he were told that they were wild birds, would certainly, I think, be ranked by him as well-defined species. Moreover, I do not believe that any ornithologist would place the English carrier, the short-faced tumbler, the runt, the barb pouter and fantail in the same genus.[6]

A few examples of these true-breeding pigeons are shown in figure 3.1. The grown-up chicks of true-breeding pigeons looked like their parents and acted in the same manner, much as the progeny of tigers resembled their parents rather than leopards, even though tigers and leopards are related species. If you had seen the actual birds in flight or at rest, you too would think that you were seeing different species of birds. But Darwin instead thought that the common ancestor of every true-breeding fancy pigeon was the wild rock pigeon, *Columbia livia*. The true-breeding, different heritable characteristics instead were the result of "artificial"—human-directed—selection over many generations.

Darwin tested his hypothesis. If the very different characteristics of each hypothetical true-breeding species were the result of artificial selection guided by humans—pigeon breeders systematically mating individual pigeons who had a particular variation present in the wild rock pigeon population, then restoring the range of variation should yield rock pigeons. His experimental procedure was simple; the pigeons were allowed to mate freely. The result of Darwin's "reverse-breeding" experiment—deliberately mixing and mating true-breeding pigeons—was rock pigeons. The members of the London pigeon clubs must have been horrified when they

Mr. Esquilant's Short-faced Baldheads.   Mr. W. Smith's White Pouters.   Mr. Wicking's Jacobin, Magpie, and Swallow.
Mr. Harran's Carrier Cock.   Mr. Wicking's Magpie and Jacobin.
Mr. Harrison Weir's White Fantails.   Mr. Wicking's Brunswick and Nun.   Mr. Percival's Turbit.
PRIZE PIGEONS AT THE SHOW OF THE PHILO-PERISTERON SOCIETY, RECENTLY HELD IN FREEMASONS' HALL.

FIGURE 3.1 True-breeding pigeons

In Charles Darwin's time, pigeon fanciers were convinced that their true-breeding pigeons were the descendants of different species because when mated, the offspring looked and acted like their parents. Darwin's pigeon-breeding experiment tested this premise. He mated the different true-breeding pigeon breeds, restoring the pool of variation. The result was the common ancestor of all of the true-breeding fancy pigeons—the dull rock pigeons that now inhabit cities.

discovered the reason behind Charles Darwin's interest in pigeon breeding.

In short, Darwin showed that the true-breeding pigeons had a common ancestor. His pigeon-breeding experiments and analyses of the findings of embryology reinforced his belief that the distinction between a species and a variety is very fuzzy. That position is explicit in the concluding chapter of *On the Origin of Species*, where Darwin states that "the only distinction between species and well-marked varieties is, that the latter are known, or believed, to be connected at the present day by intermediate gradations, whereas species were formerly thus connected."[7]

EPIGENETICS AND FOOD

In the first and subsequent editions of *On the Origin of Species*, Charles Darwin adopted his grandfather Erasmus's premise that heritable features can derive from the direct influence of the environment, and he gathered data to support this claim.

Francis Darwin also pointed out that his father had "one quality of mind which seemed to be of special and extreme importance in leading him to make discoveries. It was the power of never letting exceptions pass unnoticed."[8] Charles Darwin systematically sought out "exceptions" that supported Erasmus's theory. Most exceptions came from far afield. The penny post, introduced on January 10, 1840, was Darwin's window on the world. All that you had to do to get in touch with almost anyone was to buy and paste a stamp on an envelope and place it in the mail. Darwin took to the post, penning, mailing, and receiving a torrent of letters. More than fourteen thousand letters have survived. His biographers believe just as many have been lost. The exceptions mounted up. The stream of letters Darwin received described, among other things, birds that had lost the power of flight, the hundreds of species of wingless beetles living on the island of Madeira, and blind rats in the caves of Kentucky.

On the Origin of Species specifically discusses the direct effects of the environment on speciation. Though Darwin concludes that natural selection might have acted to preserve beetles living in Madeira who had lost their wings from blowing off to sea, he could not envision any selective advantage to losing the power of vision, even in cave-dwelling animals in the "deeper and darker recesses of the Kentucky caves."[9] Darwin thus attributed losing vision "wholly to disuse." Blind cave-dwelling animals continue to be discovered. The blind, waterfall-climbing cave fish *Cryptotora thamicola* has been found in Thailand. It lives in caves with streams, and it climbs up fast-flowing waterfalls and cliffs by using its two front and two back fins. It's a distant relative of the goldfish.[10]

In the first edition of *On the Origin of Species*, Darwin concluded that though natural selection is the primary driving force for the evolution of species, some heritable effects can be attributed to the direct effects of the environment, which he states in unequivocally clear language: "We must not forget that climate, food, &C, probably produce some slight and direct effect."[11]

It's no surprise that what you eat affects your health. Scores of books promote diets that make you stronger, improve your looks, or increase your lifespan. Should you eat saturated fats? Is olive oil a better choice? It's clear that obesity increases the risks of diabetes and cardiovascular diseases, but should you starve yourself because that increases the lifespan of laboratory mice?

WHAT YOU EAT

It's now becoming evident that Darwin's surmise that food "probably can produce some slight and direct effect" was correct. What and how much you eat affects your health and in some instances can affect your children's health and perhaps *their* children's health. It is impossible to conduct experiments that drastically limit the diets of large populations. Epidemiologists, therefore, have turned to "experiments in history" to test these theories—instances where the supply of food was severely limited by famines, wars, or rationing. Some of the findings are not surprising. During World War II, when sugar was rationed in the United Kingdom, children's teeth had fewer cavities. Their average height also increased, thanks to the balanced diet mandated by the rationing board. However, when food is really scarce, as happens in Nepal, India, and elsewhere, children grow up stunted and prone to disease. The obvious response is to feed the children.

Anders Forsdahl, a professor of family medicine at the University of Tromso, studied conditions in Norway's impoverished northeastern Finnmark region.[12] Life was once hard, and many children were malnourished. However, when they were suddenly provided with plentiful food, there was an unexpected consequence. Forsdahl discovered that as adults, these children had a higher risk of developing arteriosclerotic heart disease. But this finding, though puzzling, didn't require an alteration to accepted evolutionary theory. The effect, like, for example, exposure to pollutants, was what you might expect from any other environmental factor acting on the "phenotype." This effect wouldn't be heritable—whatever happened to these children wouldn't be passed on to their children. However, a Swedish specialist in nutrition, Lars Bygren, who had been following Forsdahl's research, had a hunch that there might be more to the story.

The circumstances of life in Bygren's ancestral village, Overkalix, in Sweden's far northern forests, provided the basis for an "experiment in nature" that might determine whether sudden changes in diet had heritable consequences—thus demonstrating direct effects of the environment on evolution. The ancestry of the inhabitants of Overkalix had scarcely changed since the late fifteenth century. The meticulous record of births, marriages, and deaths confirmed its genetic isolation—providing an approximation to the inbred, genetically similar mice that are typically used in laboratory experiments. Until railroads arrived well into the twentieth century, Overkalix was isolated from southern Sweden during its long winters; the rough roads were impassable, and North Sea ice cut off access by water. Hunting, fishing, and the local harvest were the only sources of food.

The record of harvests throughout the nineteenth century revealed years when there were adequate harvests, poor harvests, or abundant harvests. After a poor harvest, food would start to run out by early spring because it was impossible to bring anything in from southern Sweden. After a bumper harvest, it was almost impossible to send surplus food out to sell, so everyone feasted and gorged. The result was a continual and extreme nutritional cycle—deprivation followed by abundance or abundance followed by deprivation. Bygren's initial study of the effects of the feast-famine cycle focused on the grandsons of men who were born about 1905. Astonishingly, the grandsons whose grandfathers had experienced a feast season when they were entering puberty died on average about thirty-two years earlier than their cohorts, when Bygren took into account the socioeconomic factors that impinge on lifespan.

When Bygren submitted his study for publication, it was repeatedly rejected. His data suggested a rapid, direct, heritable effect of the environment on an entire population, contradicting the "standard" synthetic theory of evolution developed during the twentieth century—environmental effects could not be inherited; Bygren surely was wrong. Bygren's paper finally was accepted for publication in 2006, after other independent studies began to suggest direct heritable effects of the environment. Bygren then extended the database to the entire nineteenth century to take account of multiple years of feast or famine and track both men and women. His data showed that the granddaughters of women who had been born or carried by their pregnant mothers during a time of food scarcity also were more likely to die earlier. In a study published in 2014, Brygen and his colleagues found that women whose paternal grandmothers had lived through a sharp change in food availability in one year—from plenty to scarcity or the reverse—also had an increased risk of dying from cardiovascular disease.[13]

The DNA code in the double helix described by James Watson, Francis Crick, and Rosalind Franklin specifies the biological instructions that interact with the environment and ultimately translate into the organs of the body and brain. (The Nobel committee does not award prizes posthumously, and Franklin died of cancer four years before the Nobel Prize was awarded to Watson and Crick in 1953.) The instruction set in the double helix includes the structural genes, which determine whether you have blue eyes or brown eyes, your susceptibility to certain diseases, and your potential height and body mass. DNA also transmits other information: transcriptional factors and epigenetic information. Structural genes do not directly translate into the bones, soft tissue, and structures of your body and brain; instead, complex intermediary processes take place. The structural information coded in the double helix of DNA must be "transcribed"—rewritten into another biological code—RNA. The structural genes are transcribed by other genes, "transcription factors," into proteins—the organic chemicals that control the biochemical processes that ultimately form living organisms. Some transcription factors are located on the DNA chain near the genes that they act on; others are far removed. And transcription factors, like the structural genes they act on, are subject to mutation and natural selection. The process also involves "epigenetic" information coded on segments of the DNA that do not in themselves specify genes. Some epigenetic factors can be heritably modified by the environment. The epigenetic information coded in these segments of DNA include "promotors," "enhancers," and "silencers" that determine when during the development of an organism a gene is "expressed," that is, when it is activated or turned off. The process by which information coded in DNA forms a living organism is far more complex than anyone had imagined until the closing years of the twentieth century.

The environmental changes in nutrition that Lars Bygren and his colleagues documented acted on epigenetic segments of the DNA code. The odd effects linked only to paternal grandmothers most likely are attributable to food-supply fluctuations modifying epigenetic information coded on the female X chromosomes. Chromosomes bundle the genes of our fathers and mothers. Females have two X chromosomes, males an X and a Y chromosome. The epigenetic feast-famine effects coded in a grandmother's X chromosome apparently were erased when the maternally transmitted epigenetic information modified by feasts or famines was paired in the second generation with unaltered information from another woman's X chromosome.

The study of epigenetic effects has opened up new insights on the mechanisms that govern evolution. The distinctions that mark different species derive from mutations on epigenetic information as well as on structural genes and transcription factors. For example, depending on how similarities are defined, humans and mice share 92 to 97 percent of their structural genes. Only 2 percent of the structural genes of chimpanzees and humans are different. Fruit flies share at least 44 percent of their genes with humans.

Epigenetic promotors, silencers, and enhancers triggered by transcription factors turn genes on and off during embryonic and later development to shape a living organism. Epigenetic events play a major role in the evolution of species, and some are directly influenced by the environment and experiences that a living organism encounters.

Epigenetic information can modify the bones, muscles, and other soft tissue of the body. Environmentally modified epigenetic information also can act on brains and modify behavior. Comparison of the epigenetic promotors and enhancers of humans, rhesus macaques, and mice during the development of the cortex shows early epigenetically induced differences in humans during the first twelve weeks of gestation.[14] Research on laboratory mice has demonstrated transgenerational, environmentally induced epigenetic modification of their behavior. The female pups of mouse mothers that lick them in turn lick their pups and are calmer.

Epigenetic effects on behaviors that transcend licking may act on monkeys and humans. Jerome Kagan, over the course of decades at Harvard University, has studied the cognitive and emotional development of hundreds of children. By the age of three years, it is evident that some children are introverted and timid while others are extroverted and risk taking. The children often mirror their parents, and Kagan concluded that this characteristic is about 80 percent inheritable. The degree of introversion/shyness was assessed in a standardized, formal "test." A child with his or her mother and a graduate student (usually a young woman) were together in a room stocked with a variety of age-appropriate toys. The mother and graduate student chatted while the child roamed about and played with the toys, often including the graduate student in play. After a while, a "spaceman" covered head to foot in a silver Mylar suit entered the room. The shyest children would immediately run to shelter between their mothers' legs and cry. The most outgoing, extroverted children ran to and climbed up the "spaceman," removing her Mylar head covering. Two years later, in another experiment, the children were subjected to very mild task-induced stress while physiologic measures of stress were monitored. The children's heart rates, respiratory activity, cortisol levels, skin conductivity, and pupil dilation, all known physiologic measures of stress, were recorded, along with a speech parameter that tracks stress levels. The stress-inducing task (approved by Harvard's human subjects' research board), consisted of first asking each child to name three characters from *Sesame Street*, a children's TV show that they had been watching for at least a year. The subject child then was asked to name six characters, then finally asked to name the character who had chased a squirrel. An entire order of magnitude distinguished the shyest and most outgoing children's physiologic measures of stress.[15]

The children's dispositions appeared to mirror their parents, but the open question was the degree to which their home environment versus their parents' genetic disposition played a role in determining whether a child was outgoing or timid and shy. If it were possible to switch children and parents, placing an introverted child in the care of extroverted parents, you might more easily get an answer, but that cannot be done. The nurture-versus-nature question was addressed at the National Institutes of Health's Poolesville monkey-research facility. There Steven Suomi had observed similar behavior in monkey infants. Timid monkey infants hid between their mothers' legs when strangers approached, while outgoing monkeys were inquisitive and forward. The monkey parents' dispositions again matched their infants. When timid infant monkeys were placed with outgoing mothers or outgoing infant monkeys with timid monkey mothers, there was about a 40 percent shift in monkey behavior after a year. Whether the slight environmentally induced shifts in behavior are heritable is an open question.

It also became apparent at Poolesville why introvert/extrovert variation in behavior is preserved in both human and monkey populations. It is sometimes a selective advantage to be extroverted and adventurous, but in other circumstances being introverted enhances survival. The adult Poolesville monkeys have the run of a fenced five-acre outdoor tract. At one point, a chain-link fence separates the enclosure from a service road. A hole in the fence was discovered, and some extroverted monkeys crept through to explore the outside world. They were hit by a truck. The driver, who didn't expect to find any monkeys on the road, could not stop in time. In this instance, being introverted and timid was a selective advantage. But I would not be writing this sentence if my grandparents had been too timid to leave Poland for a life in an unknown land and a new language. They, and I, probably would have ended up in a Nazi death camp.

The full range and extent of direct environmental influence, the degree and the time depth—the number of generations in which the environment can directly produce heritable changes in morphology—remain open questions. Ongoing research projects are looking at connections between epigenetic factors influenced by environmental conditions and first- and second-language learning, as well as at autism, obsessive-compulsive disorders, and other issues. Heritable changes linked to the direct effects of the environment have been observed in honeybees, fruit flies, fungus, and other organisms.[16] Studies such as Bygren's show that some effects can persist through at least three generations, but determining the full range and permanence of heritable changes directly linked to the environment in humans and other mammals necessarily will take decades.

Darwin's "error" of attributing a role to the direct effects of the environment was criticized throughout the twentieth century by evolutionary biologists. However, the theory he proposed in 1859, which "borrowed" his grandfather's proposal that the direct effects of the environment play a role in evolution, is now driving current research.

EVO-DEVO

As he brought *On the Origin of Species* to its conclusion, Charles Darwin finally came to the question that must have been on every reader's mind: what about humans? He once more followed in his grandfather's footsteps when he stated that "probably all organic beings which have ever lived on this earth have descended from some one primordial form, into which life was first breathed."[17]

Charles, however, differed with his grandfather—no outside force or deity governed the course of evolution. The word "all" denied any special status for humanity, and he forcefully asserted that there is no overreaching plan governing evolution. "It is so easy to hide our ignorance under such expressions as the 'plan of creation,' 'unity of design,' &c., and to think that we give an explanation."[18]

The mechanisms that Darwin proposed to explore the "origin of man and his history" further involved a line of inquiry suggested by Erasmus, a line remarkably close to current "Evo-Devo" studies that focus on the expression during ontogenetic development of genes, transcription factors, and epigenetic information.

Darwin pointed out that

the framework of bones being the same in the hand of a man, wing of a bat, fin of the porpoise, and leg of the horse,—the same number of vertebrae forming the neck of the giraffe and of the elephant,—and innumerable other such facts, at once explain themselves on the theory of descent with slow and successive modifications. The similarity of pattern in the wing and leg of a bat, though used for such different purpose,—in the jaws and legs of a crab,—in the petals, stamens, and pistils of a flower, is likewise intelligible on the view of gradual modification of parts or organs, which were alike in the early progenitor of each class. On the principle of successive variations not always supervening at an early age, and being inherited at a corresponding not early period of life, we can clearly see why the embryos of mammals, birds, reptiles, and fish should be so closely alike, and should be so unlike the adult forms.[19]

You don't look much like a fish. The viewers of the 2016 PBS documentary *Your Inner Fish* might have been surprised to learn that the human embryo at one stage has features retained in fish. However, both Erasmus and Charles realized, these features are not retained in the arc of your ontogenetic development. The "why-and-how" answer involves epigenetic information and transcriptional genes that turn structural genes on and off and determine what they do.

MISSING EVIDENCE

Neither Erasmus nor Charles took account of the "hard" fossil evidence of hominin evolution, because none then had been identified. However, Charles correctly surmised one of the behavioral consequences now revealed by the fossil record and comparative studies of living apes—the relation that holds between standing and walking upright and tool use.

Hominins are defined as beings in, or close to, the line of human descent. The fossil skull found in 1856 in a limestone quarry in Germany's Neander valley wasn't recognized as belonging to an extinct Neanderthal hominin until 1886. One of the anatomists who had examined it thought that it belonged to a Russian Cossack who had died during the Napoleonic wars. Today, specimens of fossil hominins are continually being unearthed. The discoveries are often featured prominently in press releases and TV productions as well as in papers in scholarly journals.

One of the oldest, perhaps the oldest, hominin fossils yet found was excavated in Africa in Chad from the sands of the Sahara. The hominin lived about seven million years ago, close to the time when hominins and apes split from our common ancestor. The desert then was a wooded area with lakes and hosted abundant life. However, only a cranium and a few teeth were found, so it is difficult to make any inferences about the hominin's lifestyle. Skulls usually are the focus of interest because they show that hominin brain size has increased over the past seven million years. Since brains use lots of energy, the logical inference is that larger brains signify lifestyles that are more complex and advanced than those of present-day apes.

The fossil record also directly bears on the evolution of a singular human characteristic: upright bipedal locomotion—being able to walk, run, and stand upright for protracted periods. The oldest hominin fossil that shows a transition from arboreal to terrestrial life and concurrent upright bipedal locomotion dates back 4.5 to 4.3 million years. The fossil species, named *Ardipithecus ramidus*, "Ardi," was discovered in 1992 in Ethiopia by a team led by Tim White. The shape of Ardi's pelvis and foot suggests changes that aided walking and standing upright.

Earlier, in 1974, a fossil that lived about 3.2 million years ago, bearing the species name *Australopithicus afarensis*, was found in Ethiopia by a team headed by Don Johanson. Johanson named the fossil Lucy. Lucy and her relatives definitely could walk upright. Tim White, working with Mary Leakey in 1978 in Tanzania, found other *afarensis* bones that are about a half-million years older. They also found fifty-nine footprints preserved in hardened volcanic ash of two adults and a child strolling together. The footprints are not very different from those you and your child might make when walking on wet sand. Australopithicines, including Lucy, had brains that were barely larger than those of chimpanzees, though their bodies were adapted to walking and standing upright. However, they were better at climbing trees and not as adept at walking as *Homo habilis*, who lived about two million years ago.

Louis and Mary Leakey in 1960 dubbed the *Homo habilis* fossil of a twelve- or thirteen-year-old male that they found in the Olduvai Gorge of Tanzania "Handy Man" because stone tools made by flaking chips off rocks, "Oldowan tools," were found at the site. Other fossils that are similar in varying degrees to Handy Man have since been found. Their wrists, ankles, and hands show that they were well adapted for walking and had hands that, like ours, had a "precision grip." They could easily grab and hold on to things. Their brains were about 50 percent larger than earlier Australopithecine fossils. Whether *Homo habilis* was the direct ancestor of modern humans, as Louis Leakey believed, or later *Homo erectus* fossil hominins or other fossils were our direct ancestors is subject to intense debate. The seemingly endless claims of fossil hunters to have unearthed a "new" species or the "missing link" often are based on very small differences.

In virtually every conference on human evolution, diagrams are displayed that look like a tree or bush sprouting out branches as it climbs to the sky. Each branch might bear a Latinate name of a different hypothetical hominin species. Five percent differences in brain size, a smaller difference than that would be evident if you were to look around the conference room and observe the heads of the scholars in attendance, are often the basis for a claim that a new, hitherto unknown hominin species has been discovered. Fossils found in 2013 in a site at Dmanisi in Georgia have brain sizes that vary from 546 cc to 780 cc. If the fossil skulls had been collected at different sites, most paleoanthropologists would have concluded that they represented different species, but they all seem to have lived and died at the same time in the same place. The range of absolute variation in brain size is smaller than you'll find in people living in most cities today. Charles Darwin's counsel, repeated throughout *On the Origin of Species* on the distinction between a variety and a species, remains relevant: "The only distinction between species and well-marked varieties is, that the latter are known, or believed, to be connected at the present day by intermediate gradations, whereas species were formerly thus connected. . . . We shall at least be free from the vain search for the undiscovered and undiscoverable essence of the term species." With apparent amusement, Darwin went on to assure his readers that "the endless disputes whether or not some fifty species of British brambles are true species will cease."[20]

The press releases announcing the discovery of a new species almost never take into account the range of human variation that would be apparent in the pressroom.

MEALS AND TOOLS

Darwin apparently didn't think much about what was on the menu. As pointed out earlier, had he spent more time in the Galapagos, he might have discovered that natural selection can act swiftly. The beaks of the Galapagos finches rapidly change in response to the food supply. When the climate shifts to favor hard nuts, beaks became shorter and stronger. When the climate shifts to favor softer nuts, beaks became longer, allowing the birds to scoop out more food.[21] The hominin fossil record reveals connections between morphology, behavior, and food.

Daniel Lieberman in his 2013 book *The Story of the Human Body* explored these issues. Walking upright makes it possible to see things, including plants and animal carcasses, that you would miss if you were hunched over on all four limbs—a selective advantage in the "struggle for existence" for bipedal-capable hominins four million years ago as they moved through forested areas. Bipedal walking also consumes far less energy than moving on all four limbs. Although the hominins who lived four or five million years ago were not optimally configured for walking, the skeletal evidence suggests that they were able to roam over distances that far exceed the two-kilometer range of present-day chimpanzees. The later *Homo erectus* anatomical complex—legs, pelvic structures, feet, and muscles including the gluteus maximus (the "butt") that work as a spring when running—allowed hominins to hunt by running down animals.

A no less valuable advantage to upright posture is being able to use tools. In his 1871 book *The Descent of Man*, Charles Darwin stressed the link between upright posture and tool use. *The Descent of Man* is largely based on material that he had gathered before the publication of *On the Origin of Species* and on "commonsense" observations (some of which didn't turn out to be sensible).[22] Darwin pointed out the obvious fact that if you can stand and walk upright instead of resting or moving on all four limbs, you can use your hands for all manner of other things. Freeing hominin hands to use and make tools set the stage for further adaptations that enhanced the value of tools and of being able to make them.

CHIMPANZEE, THE TOOL MAKER

That raises the question of when and who first made and used tools. One of the titles for essays on human uniqueness used to be "Man, the Tool-Maker" or "Man, the Tool-User," but an alternative title for a different point of view could be "Chimpanzee, the Tool Maker and User." In the 1970s, Jane Goodall's research team showed that the chimpanzees of the Gombe Reserve in Tanzania used and made tools. Their tools were sticks and leaves; they are not the stone tools that paleoarcheologists necessarily focus on. Sticks and leaves don't survive for millions of years. For that matter, few of the tools that you or I use would survive for several million years, either.

Leaves served as napkins and as sponges for drinking. They stripped small branches to make termite-fishing sticks. When chimpanzees walk upright for short periods of time, sticks become weapons. Faben, a dominant "alpha" male, even waved a stick to threaten his reflection in a mirror.[23] Trash discarded by Goodall's research team was opportunistically used. Mike, a chimpanzee who wasn't especially strong, achieved alpha-male status by charging and clattering two or even three empty kerosene cans before him. Different patterns of tool use mark chimpanzees living in different locales. In West Africa's Ivory Coast, chimpanzees living in the Tai rain forest used stones and wood clubs to open nuts. In the season in which hard nuts were plentiful, they were pounded open with stone hammers. During the soft-nut season, wood clubs were used.

Hedwige and Christophe Boesch, in their decade-long study that began in 1979, observed and filmed chimpanzees gathering nuts on the ground, then looking around for a suitable hammer and anvil to crack them open. When the chimpanzees climbed up a tree to gather and crack nuts, they carried a suitable hammer, a wooden or stone club, with them. Tree branches served as the anvils. Female chimpanzees devoted more time to nut cracking and were more proficient at it. Chimpanzee mothers also acted as mentors to their young, guiding them through the process of placing a nut on the anvil so it didn't roll off. If mother and child were sitting high up in a tree, the mother would stow the hammer so that it didn't fall to the ground. The films show mothers patiently correcting their pupils' errors.

Stones were scarce in the Tai rain forest. The chimpanzees' hammer stones, which can be identified by the strike marks on them, were stored for use during the next harvest season. In 2006, Julio Mercator, Christophe Boesch, and their colleagues discovered a prehistoric "Chimpanzee Stone Age." Hammer stones bearing strike marks were found that had been stored by nut-cracking chimpanzees as long as 4,300 years ago. No human farming villages then existed in the Tai forest, precluding the chimpanzees from acquiring the technique by imitating humans. Some genius chimpanzee probably invented the technique, which had been transmitted from one generation of chimps to the next over a span of time almost as long as that of written human history. Chimpanzee cultures differ with respect to tool use and diet. In Western Uganda's Budango Forest Reserve, they use leaves as sponges to drink from clay-rich water holes under trees, and eat clay from clay pits, selecting clay that has a lighter color than the forest's red-brown floor.[24] They also eat the soil of termite mounds. If the hominins who lived during the transition from arboreal to terrestrial life were as enterprising as chimpanzees, they would have been using sticks and leaves along with stones.

Homo habilis, a.k.a. "Handy Man," wasn't the first hominin species who made stone tools. Stone tools crafted by striking stones along fracture lines, in much the same manner as how the facets of gemstone diamonds are "cut," were made 700,000 years earlier—3.3 million years ago in West Turkana, Kenya.[25] Charles Darwin lacked both the evidence of the hominin fossil record and comparative evidence on tool-using apes, but he was correct in linking upright bipedal locomotion and tool use.

MOTOR CONTROL

It's amazing to view the Tai chimpanzees cracking nuts, but they are clumsy and slow. Chimpanzees rarely hit their targets when they hurl sticks or other objects. One of the most striking observations of the limits of chimpanzee hand-eye coordination can be viewed on YouTube. It was filmed in the 1990s and shows Kanzi, the bonobo (pygmy chimpanzee), studied by Sue Savage-Rumbaugh, attempting to drive a golf cart. The cart repeatedly lurches into the bushes off the path. Bonobos are by some primatologists thought to be a distinct chimpanzee species, different from the common chimpanzee, *Pan troglodytes*. When raised from infancy in a human, language-using environment, bonobos and chimpanzees can use nonvocal channels to communicate using words and can understand sentences that have simple syntax.[26] The Gardner project's chimpanzees used American Sign Language to communicate what they wanted, what they feared, and to convey their opinions. However, though there have been many attempts to teach apes, usually chimpanzees, to talk, they cannot learn or execute the motor acts that underlie speech. La Mettrie, the author of *L'homme machine*, a materialist tract published in 1747, had stated that if an ape could be taught to talk, he would be a perfect gentleman, thereby disputing the existence of the soul and God. In the Soviet Union, a project aimed at teaching apes to talk was initiated, with the apparent objective of proving that God did not exist.

The studies that my colleagues and I published in the 1970s, which showed that apes did not have an adult-like human tongue and supralaryngeal vocal tract, were misinterpreted. Their anatomy did not preclude their being able to talk. As we pointed out, apes could have produced all aspects of speech except for the quantal vowels and some consonants—if their brains had allowed them to learn and execute the complex voluntary motor acts involved in talking. Apes also seem to be incapable of learning to execute other motor tasks. Has anyone ever been able to teach a chimpanzee to dance?

MOTOR CONTROL AND CREATIVITY

When the *Beagle* dropped anchor in Tierra del Fuego, Charles Darwin most likely saw Oldowan tools in use that were similar to ones made 3.3 million years ago. Stone tools are indices that provide insights on the evolution of the neural capacity for fine motor control and creativity, which I and others believe are linked.

Mastering the art of fabricating Oldowan tools appears to involve having the capability to execute visually directed hammer strikes *and* having some form of language. Nicholas Toth and Kathy Schick work together at the Stone Age Institute of Indiana University, where they have attempted to approximate the stone-working techniques used at the sites at which stone tools were found. In the course of Sue Savage-Rumbaugh's bonobo project, Nicholas Toth tried to teach the bonobo Kanzi to make an Oldowan tool. While Kanzi watched, Nicholas Toth hammered at a suitable stone along a fracture line. Success in making an Oldowan stone tool entails striking a "core" stone along a fracture line, shearing off flakes. The sharp edges of the stone flakes as well as the worked-on core can be used as cutting tools. Though Nicolas repeatedly tried to convey the technique in deliberate slow motion, Kanzi never discerned the point of the lesson, which was to hit the stone along a fracture line, shearing off flakes. Instead, a frustrated Kanzi eventually hurled a stone blank against the concrete floor of the cage they were sitting in.

The Oldowan tools recovered by archaeologists at hominin sites in Africa, moreover, show signs of retouching—resetting the edges

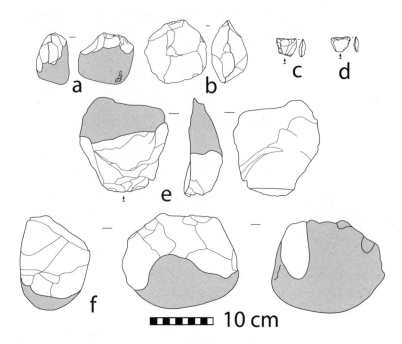

FIGURE 3.2 Oldowan tools

a–d. 1.8-million-year-old Oldowan tools from the Olduvai Gorge, Tanzania.

a: chopper; b: core; c and d: flakes that had been retouched after use.

Source: Redrawn after M. D. Leakey, *Olduvai Gorge: Excavations in Beds I and II, 1960–1963* (Cambridge: Cambridge University Press, 1971).

e–f. 3.4-million-year-old tools from Lomekwi 3.

e: unretouched flake; f: chopper.

Source: Redrawn after S. Harmand et al., "3.3-Million-Year-Old Stone Tools from Lomekwi 3, West Turkana, Kenya," *Nature* 521 (2015): 310–315. Courtesy John Shea.

after the tools were used. Cut marks on animal bones confirm how the tools were used.[27] Making an Oldowan stone tool requires hammering on a fault line. Kanzi didn't get it, but some hominins as far back as 3.3 million years ago had mastered the Oldowan technique. Perhaps learning to make an Oldowan tool entails the instructor

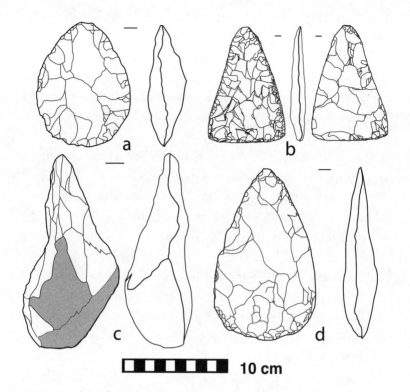

FIGURE 3.3 Acheulean hand axes

a–b. Later Acheulean hand axes from Azraq Lion Spring (Jordan) and Abilly (France).

c–d. Early and Middle Pleistocene handaxes from Ubeidiya (Israel) and Maayan Barukh (Israel).

Source: Redrawn after F. Bordes, *Typologie du Paléolithique ancien et moyen* (Bordeaux: Delmas, 1961); J. Shea, *Stone Tools in the Paleolithic and Neolithic of the Near East: A Guide* (New York: Cambridge University Press, 2013). Courtesy John Shea.

and pupil sharing information through the medium of language. This premise could be tested. If you cannot teach modern humans who don't understand your language to master Oldowan tool making, then we could infer the presence of linguistic ability at a level that exceeds that of present-day chimpanzees in the hominins who produced these tools 3.3 million years ago.

To most people, Oldowan tools don't look like tools. If you were a casual visitor at an archaeological site, you might think that you were looking at a battered stone instead of a tool that someone had deliberately made. However, virtually anyone looking at an Acheulean hand axe realizes that it is a tool.

The name Acheulean derives from the city where these stone tools were first found—Saint-Acheul, a suburb of the city of Amiens, in northern France. The tourist brochure for Amiens, which is on the river Somme, notes its cathedral and that it was the home of Jules Verne. The Jardin Archeologique, where the hand axes were found in 1872, can also be visited. The oldest Acheulean hand axes are about 1.76 million years old. The most recent hand axe (other than ones made by archaeologist "reenactors") was made 300,000 to 200,000 years ago. They most likely were used to scrape and cut skins, meat, and bones. It is difficult to imagine using one as an axe to chop down a tree.

Several hundred thousand Acheulean hand axes have been found throughout Eurasia. They are the signature artifact associated with *Homo erectus*, but though their shapes somewhat differ—some wider, others elongated—over the span of about 100,000 generations, they all essentially look similar. One hand axe looks much the same as another—but this similarity is puzzling. No one seems to have thought of thinking outside of the box to produce an improved tool, which is evidence of an absence of the impulse to create something new, an impulse that marks modern human beings. It has generally been assumed that the technique for making a hand axe was culturally transmitted, someone instructing a novice or a novice copying an expert. But the "frozen" form, the absence of innovation, is noteworthy. Why do they all look the same, for such an extended span of time? Some paleoarcheologists are beginning to think that the instruction set for making an Acheulian hand axe might have been genetically transmitted. In other words, no one had to *learn* anything.[28] Making a hand axe would have involved releasing a genetic instruction set, a "template," in much the same manner as how birds make nests or spiders spin webs.

The genetic-program proposal is controversial. It's more likely that a profound leap in hominin cognitive capabilities occurred about 500,000 years ago. The next chapter will discuss recent advances in understanding the neural circuits that confer human cognitive ability, including some of the transcriptional genes and epigenetic factors that shape the human brain. Studies that in essence constitute genetic "time travel," recovering and interpreting DNA from long-dead hominins, suggest that cognitive flexibility was enhanced in our immediate ancestors and their extinct cousins, Neanderthals and Denisovans, to a degree that encouraged experimentation and new forms of activity—including tool making.

BAKING A CAKE: LEVALLOISIAN TOOL MAKING

That proposal is consistent with the archaeological record—stone tool-making technology jumped forward about 500,000 years ago. The "Mousterian" culture associated with Neanderthals produced stone tools using the Levalloisian technique. (The names again refer to sites in France.) Making an Acheulian hand axe is akin to whittling. You start with the raw material and chip and chip. You perform much the same act again and again. The path to the final goal is always in sight. No change in the toolmaker's mindset is called for—you are on autopilot.

A different storyline is necessary to describe making a Levalloisian stone, a storyline with a twist. The toolmaker must first chip away at a large stone, but the objective is to prepare a core, which doesn't at all resemble a tool. The prepared stone core best resembles a headless, footless, stone turtle. The next step entails a conceptual shift. When you have decided that the turtle-like stone core looks right, you have to switch to a different technique. You can either apply pressure to the core with a stick or strike it briskly with a hammer stone in such a way that a finished tool fractures off from the core along a fault line.

The Levalloisian technique is conceptually similar to baking a cake. The initial steps in cake making involve mixing and stirring flour, milk, yeast sugar, etc., into a thick paste. The gooey paste has no resemblance whatsoever to a cake. It doesn't look like a cake, doesn't taste like a cake, doesn't feel like a cake. No sensory information could lead you to believe that it had any relation to a cake. But after shaping the dough, you then abruptly switch to a different technology, one that entails knowledge of entirely different parameters.

The mass of dough is put into a fired oven. First you must have an oven—you or someone must have previously made an oven—a device that does not have anything to do with chipping stones. The oven could have been made using clay, or stones could have been used with the clay. Fuel must have been gathered, and a fire must have been started. If you know the correct time and have some sense of how to get the right temperature, you'll put the gooey mixture into the oven. In a little while you will have a cake, if you take the cake out at the right time, which depends on the temperature and ingredients—if you don't have a thermometer, which also entails a fund of prior knowledge based on direct experience or information conveyed through the medium of language.

The Levalloisian stone tool–making technique doesn't involve as great a conceptual jump as does making a cake, but it entails having to jump mentally from chipping away at a core to applying pressure or percussively flaking off tools. Moreover, it entails knowing that stones will precisely cleave along fault lines. The knowledge base

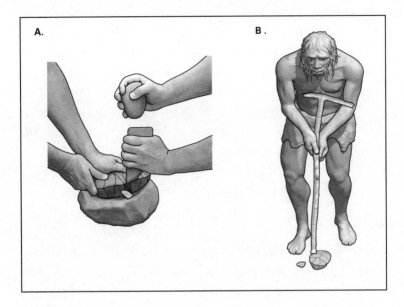

FIGURE 3.4 Levalloisian technique

Advanced Levalloisian stone tool–making technique. The toolmaker first prepares a "core" (A), positioned on an "anvil" stone. Blade tools then may be made either by striking the core (A) or by pushing on a stick (B).

Source: Redrawn from author's sketch.

and technique surely must have been transmitted through the medium of language. It is difficult to see how you could convince anyone to make a turtle-like core, a laborious task, without explaining what it will be used for.

Neanderthals survived for a long period in challenging Eurasian environments. But the variety and sophistication of the stone tools that they produced never matched those of the modern humans who displaced them. John Shea has pointed out that stone tools can be regarded as indices for increases both in the cultural complexity of human societies and in cognitive flexibility. The proliferation of specialized tools adapted to different functions—notched stone tools, hammers, awls, and blades—is a mark of cultural complexity. Different solutions to similar problems reflect cognitive flexibility— that is, creativity. Though apes and other nonhuman primates use stones as percussive tools (hammers), nonhuman primates do not make stone cutting tools. The literally hard evidence—stone tools— for the evolution of cognitive flexibility in hominins shows that the process started millions of years ago and accelerated during the Paleolithic.

ART

I have spent thousands of hours—years—engaged in a basically useless activity: taking photographs, developing and printing film in darkrooms, more recently with computers and digital printers. Virtually everyone at some point feels compelled to produce works of art, be it as simple as doodling images on the back of a paper napkin.

Art seems to be a distinctly human preoccupation. People often paint images on walls, rocks, cloth, pieces of wood, etc. Throughout recorded history, complex and labor-intensive technologies have been developed to paint decorative patterns in cloth, weave tapestries, produce photographs, etc. Some forms of art are ubiquitous across culture, both in history and prehistory. In virtually all places and times people have made and adorned themselves with stones, green stones, dark stones, shiny stones such as diamonds, glass beads, silver, gold, enameled metal—jewelry. People color parts of their body—red lips, pink cheeks, blue faces, tattoos of all sorts everywhere—faces, necks, legs, hands, feet, soles. Every visible part of the human body can and has been tattooed, colored, and adorned in some culture at some time.

Neanderthals may have produced some works of art. Some scribed marks on the rock walls in a cave in Gibraltar have been attributed to Neanderthals crafting a work of art, but the dating is extremely uncertain; the marks could have been made after Neanderthals became extinct.[29] In contrast, a "high-tech" paint factory operated by humans was in production about 100,000 years ago, in the Blombos cave, about 186 miles east of Cape Town in South Africa. In the factory, abalone seashells were used as vessels in which a mixture of ochre, bone, and charcoal was placed. The mixture was heated over a period of days and stirred using bone implements. Chemical analyses suggest that liquefied bone marrow probably was used as a thickener. The shell vessels were reused repeatedly, and other bones were used to apply the paints.[30] Red ochre has been used throughout recorded history for body painting, and traces of red ochre remain in the stone mortars in which it was ground in prehistory. Today it is still used in cosmetics—think blush and lipstick.

If humans acted then as they do today, there probably was a long period well before the paint factory came into being when red ochre powder was directly applied to the skin.

Stone tools with inscribed decorative lines made 75,000 years ago have also been found in the Blombos cave. Seashells that were pierced to make necklaces have been found at other sites in Africa dating back 124,000 years. Other jewelry appears in the form of pierced ostrich eggshell, bone necklaces, and pendants. Paintings on cave walls and rock formations have been found in Europe and Australia after part of the human African population moved into those regions. The abrupt "cultural revolution" that was attributed to the discovery of art in Western Europe 45,000 years ago reflects the discovery of things that people brought with them when they moved there from Africa.[31]

IN SHORT

Charles Darwin borrowed numerous concepts from his grandfather Erasmus, including the "error" of attributing some role in evolution to the direct effects of the environment. Charles, like Erasmus, proposed that all forms of life derived from a common ancestor and used the evidence from embryology to support this claim. However, he differed with Erasmus, who adopted the eighteenth-century deist position that a master plan governed the course of evolution. Perhaps wanting to distance himself from the speculative tone of Erasmus's book *Zoonomia*, Charles instead assiduously collected data and conducted experiments to support his claims—but his assertion that Erasmus had not influenced his work rings hollow.

The experimental design of modern science, testing a theory against data and modifying it if necessary, was novel in Charles Darwin's time. He tested the hypothesis of common ancestry through a reverse-breeding experiment in which true-breeding pigeons that were supposed to be the descendants of different species instead yielded common rock pigeons when they were allowed to mate. The pigeon-breeding experiment exemplifies how Charles Darwin differed from Erasmus Darwin, stressing data and avoiding speculation. To Charles Darwin, the distinction between a species and variety was arbitrary—the view of present-day evolutionary biology.

Darwin in the first and subsequent editions of *On the Origin of Species* claimed that in some instances natural selection was not the sole agent of evolution. The environment could produce heritable effects. It took about 150 years for this proposition to be validated, when the transgenerational effects of abrupt changes in the food supply in humans were documented. Current studies that focus on epigenetic segments of DNA and transcription factors are exploring the biological mechanisms involved. Evo-Devo studies shed light on the developmental processes that determine whether an organism develops into a fish, mouse, ape, or human,

Neither the hominin fossil record nor the archaeological record was available to Charles Darwin, but his insights concerning the relationship that holds between tools and upright bipedal locomotion and posture were correct. The archaeological record also suggests that aspects of behavior, such as art, that distinguish humans from other species by reflecting humans' enhanced creativity, appeared in the last 500,000 years or so.

4

CRAFTING THE HUMAN BRAIN

CHARLES DARWIN'S TOOLKIT

No one knows completely how brains work. For example, many studies show that when we or a mouse looks at the world, fragments of the scene—colors, shapes, and movements—are discerned and stored in different parts of the brain. How these bits and pieces are put together remains a mystery. It's as though a concert has been announced, the programs have been printed, the musicians who play the different instruments have been hired and are ready to perform, but the score and conductor are missing. The theories that currently seem to offer a possible solution to understanding how brains work and how the human brain evolved posit circuits that link activity in different parts of the brain. We are far from understanding how this may work—or how and why the human brain evolved—but the toolkit packaged in *On the Origin of Species* is proving its value in this quest.

Darwin's toolkit wasn't limited to natural selection acting on chance variations. He pointed out how the modification of an existing organ to serve a new end, what we have called "recycling," could radically alter the behavior of living organisms. Neural "organs," structures of the brain that initially may have solely regulated simple motor responses in creatures that lived before the age of the dinosaurs, have been modified to enhance their capabilities so as to allow us to talk. These neural structures include the basal ganglia, structures deep within the human brain, which also play a central role in some of the neural circuits that confer distinctively human cognitive capabilities, including being able to form and comprehend complex sentences. These neural circuits also play a part in conferring the creative abilities manifested in art since our distant precursors diverged from the common ancestor that we share with present-day apes. Darwin also observed the continuity of evolution revealed by embryology. In the examples that follow, I will discuss the findings of current studies that suggest that recycling, along with the transcriptional and epigenetic factors that govern the expression of structural genes during development, crafted the human brain. In this instance, the overused phrase "cutting edge" applies. The principles that the world of science learned of more than 150 years ago are guiding research on one of the most intractable problems of twenty-first-century research: how our brains work and how they evolved.

NEURAL CIRCUITS

A discussion on how brains work first entails showing that the traditional theory, proposed in the early years of the nineteenth century, is wrong. *Coming of Age in Samoa*, Margaret Mead's account of life there, was controversial from the day of its publication in 1928. When she spoke in 1973 at the International Congress of Anthropological and Ethnological Sciences, her depiction of sexual mores in Samoa was still being disputed. But what I recall didn't have anything to do with sex. On the stage in Chicago, Mead was bantering with Sol Tax, one of the congress's organizers. At age seventy-two, she still had an irreverent attitude toward received wisdom and declared that the assembled scholars, like most people, usually follow the "fumba." We continue to believe what our ancestors believed.

The fumba that continues to pervade neurophysiology dates back to Franz Joseph Gall, who in 1796 claimed that the human brain contained twenty-seven "organs," each determining an aspect of personality, such as piety, or of cognitive capacity, such as language. These "faculties" could be located by a skilled phrenologist by examining a person's skull, locating bumps and depressions and taking measurements. The underlying premise of phrenology is still with us, though under another name, in studies that claim that independent neural "modules"—self-contained neural systems—each confer some aspect of behavior, such as language or artistic ability or personality. The sites of the hypothetical neural centers that trigger fear or pleasure or that confer complex aspects of behavior such as language no longer are located by examining bumps on the surface of a person's skull. However, current neophrenological studies presented in books published by university presses and papers published in scholarly journals claim that the human brain is an assemblage of discrete neural structures, each conferring an aspect of cognition or language or specific aspect of behavior such as fear, pleasure, empathy, or moral conduct. Phrenology was a key feature of *Vestiges of the Natural History of Creation*. *Vestiges*, a jumble of phrenology and intelligent design published anonymously in 1844, was attacked by the clergy because it deviated from the story in Genesis. It was also attacked by the leading scientists of the day for its errors and absence of supporting data. *Vestiges* convinced Charles Darwin to delay publication of his evolutionary theory until he had an airtight case. Darwin published *On the Origin of Species* only when he received a letter from Alfred Wallace that sketched out a theory of evolution similar to the one that Darwin had working on for more than twenty years.

Robert Chambers, the anonymous author of *Vestiges of the Natural History of Creation*, was a member of the Edinburgh Phrenological

Society, founded in the 1820s by George Combe, one of the leading phrenologists of the day. Combe was an international celebrity. When John Audubon visited Edinburgh to promote his paintings of birds, one of his priorities was to be examined by Combe. After measuring Audubon's skull and moving his fingers along it to ascertain the locations and extent of the bumps and hollows, Combe doubtless concluded that Audubon possessed a highly developed "Faculty of Art."

However, in other instances, the shape of the skull and the size of the bumps and hollows did not predict the subjects' abilities or propensities. Some homicidal maniacs had large bumps that ostensibly supported the "Faculty of Piety," and some celebrated clerics had very small piety bumps. A large language bump did not always correspond to a facility with language, and so on. However, in 1860 Paul Broca resuscitated phrenology when he published a study in which he located the brain bases of talking in a specific part of the human brain: Broca's area of the cortex.

THE BROCA-WERNICKE LANGUAGE ORGAN

Broca's reasoning cannot be faulted. Every device, tool, or implement that he might have encountered was made up of discrete parts that each carried out a particular function. Complex instruments such as watches had a set of discrete parts—a spring that when wound powered the watch, the escapement that set the time interval, gears that moved the hands of the watch, and so on. You might be able to learn how a watch worked by systematically destroying each part and observing the effect on the functioning of the whole. If you destroyed the gears that drive the hands, the hands wouldn't move, though the watch would still tick. Disabling the spring would stop everything. In short, you might be able to map out the functional architecture of watches. The underlying premise of phrenology, that the human brain is an aggregation of discrete organs or faculties, was reasonable. You don't have to believe that a bump on a person's skull housed a self-contained system that controls moral acts or conferred language. Some other part of the brain might constitute the neural device that constitutes the "Faculty of Speech," while another part might regulate the process that allows people to understand distinctions in meaning conveyed by syntax.

As chapter 2 pointed out, Tyson's dissection of an orangutan in 1699 had shown that apes had much smaller brains that humans, and subsequent studies had shown long before 1860 that the outer layer of the human brain, the neocortex, was disproportionately larger than it was in apes. The neocortex thus might be the site of the faculties that distinguish humans from apes. If you approached discovering the functional architecture of the human brain in the same manner as you would a watch, you could systematically destroy parts of the cortex and observe the behavioral consequences. That was and remains a forbidden experiment, but "experiments in nature" occur when a stroke or other trauma destroys part of a person's brain, leaving the patient alive, at least for a while.

Paul Broca in 1861 published his study of the brains of two diseased patients. One patient, the fifty-one-year-old M. Leborgne, had suffered a series of neurological attacks. Leborgne could not produce any recognizable words other than an utterance that was heard as the syllable "tan"; Broca described him as patient "Tan." The right side of Tan's body also was paralyzed. Tan's inability to communicate made it impossible to determine whether he had suffered other cognitive or linguistic deficits. According to phrenological maps, the "seat" of the "Faculty of Language," the part of the brain that controlled language, should have been between Tan's eyes. Broca did not question the underlying premise of phrenology, that a specific part of the brain was the seat of language, but he thought that it was located elsewhere, and the autopsy showed damage to the left side of the front of Tan's cortex. Unfortunately, Broca limited his observations to the surface of Tan's brain rather than sectioning it to determine the full extent of the damage.

A few months later, Broca examined a second patient who after a stroke could speak only five words. The autopsy showed brain damage to approximately the same cortical area. Broca concluded that the brain's language organ was localized in this cortical area, "Broca's area." In virtually all textbooks that touch on language, most medical textbooks, articles directed at general readers, and current research published in peer-reviewed scientific journals, Broca's area is described as the location of the neural mechanisms that confer language.

The neural apparatus that confers all aspects of language is often thought to be sited in Broca's area,[1] but to Paul Broca it was the organ of the brain that controlled not language but speech. In 1874, Karl Wernicke in Germany studied a stroke patient who had difficulty comprehending speech. The patient had suffered damage to the rear (posterior temporal) region of his cortex, so Wernicke,[2] in the spirit of phrenology, decided that this area was the brain's speech-comprehension organ. Since language entails both being able to talk and being able to comprehend what is being said, Lichtheim in 1885 proposed a cortical pathway linking Broca's and Wernicke's areas. Thus, Broca's and Wernicke's cortical areas and the cortical pathway connecting them came to be viewed as the brain basis of human speech and language. Figure 4.1 shows the two traditional organs of language, one for talking, one for comprehending speech, and the pathway that connects them. Broca's "syndrome" clearly exists. It usually involves problems in both talking and comprehending the meaning of spoken and written sentences, but it never occurs when the pattern of damage is limited to the cortex.

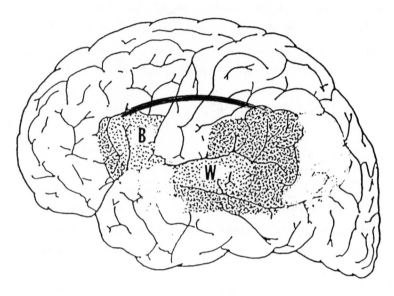

FIGURE 4.1 Broca's and Wernicke's areas and the cortical pathway connecting them

The traditional language areas of the human brain, connected by their cortical pathway. Paul Broca thought that damage to a discrete area of the cortex resulted in aphasia, the permanent loss of language. However, hundreds of CT scans and MRIs show that brain damage limited to the cortex does not result in aphasia, which instead occurs only after subcortical damage.

Paul Broca missed the true cause when he published his study in 1861.

Broca preserved the brains of his two patients in alcohol, and 150 years later, Nina Dronkers and her colleagues examined the alcohol-pickled brains using high-resolution magnetic resonance imaging (MRI). As pointed out earlier, MRIs yield three-dimensional images of the soft tissue of the entire brain. In both patients, massive damage was evident below the cortex in the basal ganglia and "white matter" that connects the neural structures that constitute the cortex. Surprisingly, the MRIs revealed that the area of the cortex commonly labeled "Broca's area" in almost every textbook or article on the neural bases of language was intact in Paul Broca's patients.

The cortical surface of the human brain has a pattern of folds, "gyruses" (ridges), and "fissures" (hollows). In the early years of the twentieth century, Korbinian Brodmann produced brain maps to delineate further the regions of the brain.[3] The human cortex has six layers of "neurons," the basic computational elements of all brains. Brodmann's microscopic studies showed that the shapes of the neurons in each of the six layers of the human cortex vary across the parts of the cortex. Brodmann's brain maps delineated different areas of the cortex in humans and other species according to the manner in which the neurons in these layers were different. He inferred that the various distributions of neurons in the layers of the cortex signified different types of brain activity, which suggested different functions and behavioral outcomes. The hunt to establish what the different Brodmann areas might be doing continues today.

The Brodmann map in figure 4.2 shows the traditional site of Broca's area in BA area 44 in the left inferior gyrus of the brain (the left lower part of the front of the cortex—BA signifies Brodmann area). However, the MRIs that Dronkers and her colleagues examined showed that BA 44 wasn't damaged in Paul Broca's patients. The cortical area in front (anterior) of BA 44 was damaged instead. But the part of the cortex that was damaged in Paul Broca's patients isn't very important. Contrary to the fumba—the story repeated over and over about Broca's area being the seat of language—damage limited to any part of the cortex doesn't result in aphasia—permanent language loss.

The results reported by Dronkers[4] were not the first to challenge the Broca-Wernicke language theory. By the end of the nineteenth century, doubts had been expressed about both the role of the cortex in language and the localization of aspects of language. In the

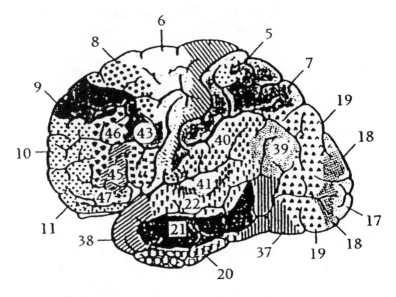

Figure 4.2 Brodmann areas of left hemisphere of the human cortex

The shapes of the neurons in the six layers of the human neocortex differ across its different regions. Broadmann hypothesized that this "cytoarchitectonic" neuronal architecture had functional consequences and assigned numbers to each area. Hundreds of studies have since attempted to determine what each area does. Area 44, Broca's area (though it wasn't actually the cortical area damaged in Paul Broca's patients), is traditionally thought to be the brain's language or speech organ.

1920s, the French neurologist Pierre Marie[5] noted that the presumed site of Broca's area often was intact in patients who had suffered a permanent loss of language. Doubts were expressed by other neurologists, but studies were limited by the need to perform a postmortem autopsy of a patient's brain. In the twentieth century, neuroimaging studies that first used computed tomography (CT) scans and then MRIs showed that the traditional Broca-Wernicke language-organ theory was wrong. Whereas autopsies of the brain are not common, the invention of CT scans in 1972 made it possible to examine thousands of brains in living patients after strokes and other trauma to the brain. It became clear that aphasia did not result from damage to any part of the cortex. It's possible that this reflects the fact that the cortex seems to be quite malleable. In blind people, cortical areas that are active in vision instead become active when they listen to speech or other sounds. However, it's clear that Broca's area is not the human brain's discrete "language organ."

NEURAL CIRCUITS

The patterns of deficits of aphasia—permanent loss of speech capabilities and other aspects of language—are real, but they do not derive from brain damage localized to Broca's or Wernicke's areas or to any other part of the cortex. In their 1986 book *The Frontal Lobes*, written for neurologists, Donald Stuss and Frank Benson stressed that aphasia never occurs absent subcortical brain damage.[6] Nor are aphasic patients' problems limited to language. The straightjacket of phrenology had put cognition and language into different "faculties," but it was evident to Kurt Goldstein, one of the foremost aphasiologists of the twentieth century, that brain damage that disrupted language also resulted in a loss of the "abstract attitude."

Goldstein, who had trained under Karl Wernicke, treated hundreds of World War I veterans. Bullet wounds had produced the classic symptoms of aphasia—problems talking and comprehending speech. But the aphasic patients also had problems in reasoning, planning ahead, and forming general principles from specific instances. They seemed to be bound to the here and now. For example, when confronted with an inkwell (in the era before fountain pens and ballpoints) that had been moved from its usual position on his desk, the patient would be at a loss as to what to do.[7] In short, the aphasic "syndrome" includes problems with speech motor control, comprehending speech and written text, and a suite of cognitive deficits.

Converging evidence from both traditional "experiments in nature" and studies that monitor activity in living subjects point to a different view of how brains work. "Neural circuits" that link local operations performed in different parts of the brain regulate most complex aspects of behavior. And a given part of the brain may perform the same or similar local operation in different neural circuits.

The experiments in nature include studies of aphasia and neuro-degenerative diseases such as Parkinson's disease, other diseases affecting the brain, oxygen deficits, and autism. The simplified circuit map in figure 4.3, for example, accounts for some of the behavioral problems associated with Parkinson's disease. Noninvasive brain-imaging techniques allow researchers to track activity in a person's brain as she or he performs a task (or does nothing). Noninvasive tracer techniques can map out the connections between parts of the human brain, its "neural circuits."

An everyday problem, one that happens thousands of times every day, can serve to convey the distinction between the local operations performed by a particular neural structure and the combined operations of a neural circuit. If your car won't start when the AAA truck arrives, the mechanic will not look for a defective "center of starting." There is no single organ or faculty that, in itself, starts your car. The usual suspect is the battery, and anyone who has any knowledge of the functional architecture of cars—that is, knowledge of how cars work—will first see if the headlights go on. That's because the battery is not a dedicated organ devoted to one and only one task, namely, starting your car. The battery also powers the lights, ignition system, fuel pump, radio, GPS, and all other electrically powered devices—including the computer that now controls the engine in most cars. If the lights go on or if the radio plays,

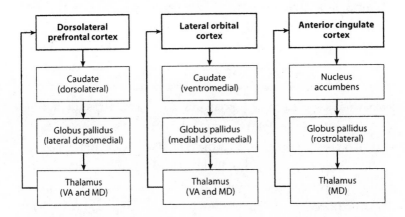

FIGURE 4.3 Some basal ganglia circuits

Three neural circuits (omitting many details) that play a part in regulating motor control, cognition, and emotion. The diagrams are based on Jeffrey Cummings's 1993 review of the behavioral deficits of patients who had suffered from strokes, coma, trauma, or Parkinson's disease. Cummings took account of the findings of tracer studies that had mapped neural circuits in monkeys and other animals. Subsequent studies on humans that used noninvasive brain-mapping techniques bear out his general conclusions. Neuroimaging studies confirm the role of the dorsolateral prefrontal circuit in a range of cognitive tasks. Lesion studies and studies of the behavioral deficits that can occur in Parkinson's disease confirm the role of the cingulate circuit in both attention and controlling laryngeal phonation. Orbofrontal circuits appear to be involved in regulating emotion.

Source: After J. L. Cummings, "Frontal-Subcortical Circuits and Human Behavior," *Archives of Neurology* 50 (1993): 873–880.

the problem instead might be in the ignition key or starting button, or in the starting relay or the starting motor—a string of devices that each perform a "local" operation and are connected in a circuit.

The battery also might be dead because the alternator that is supposed to charge it isn't working or, if your car is a gasoline-electric hybrid, because the braking-recharge system has failed. There is no discrete part in your car that, in itself, starts your car. Instead, what's involved is a circuit that links a number of parts. The battery also forms part of other circuits. As Charles Darwin might have pointed out were he with us, the battery has taken on "new" roles in the course of the "evolution" of cars. In the Subaru that I bought last year, the battery now also powers the steering system, which in older models used to be powered by a mechanical pump pushing hydrolytic fluid through a series of tubes. Batteries have continually taken on new roles throughout the evolution of cars. In the 1920s, a car's battery powered its headlights in a single circuit. Soon afterward, the battery also powered starting motors in another circuit. Shortly thereafter, the battery also powered the radio in yet another circuit. Today, in virtually all cars, the battery also powers the computer that controls the engine, the braking system, and the cruise control. Battery-powered collision-control systems soon will be standard on all cars as well.

The same part or "organ" of a car generally enters into different aspects of car "behavior." That's the case even for even basic car organs such as the engine. In conventional, stick-shift transmissions, the engine propels the car, but it also can act as a brake. The brakes in a hybrid or electrically powered car also recharge the battery. Some parts, such as the door lock, can have a single function—locking the car—but that isn't the case in cars equipped with alarm systems. To return to the issue of why your car won't start, the "behavior" starting your car rests on the entire circuit functioning. That requires not only the organs to function but the connections between them to function as well. If a wire is frayed or cut, your car won't start, even if every single component in the circuit is intact. Functioning pathways, the electrical circuits linking the parts of the car, are essential. The number of circuits in any car is minuscule compared to the neural circuits of your brain, or even the brain of a mouse. The worldwide initiative aimed at mapping out the circuits of the human brain is a multi-billion-dollar endeavor that is still in its infancy.

The fragmentary knowledge presently available points to complex aspects of human behavior such as language deriving from multiple neural circuits that link both adjacent and distant neural structures. The linked neural structures generally also are implicated in tasks and activities as diverse as motor control, memory, emotion, changing the direction of your thoughts, and finding your way home. Many of the neural organs involved have a long evolutionary history and are present in living species besides our own, but as Darwin pointed out for lungs, they have been modified to serve new roles in language. Even seemingly simple acts such as walking involve complex circuits and neural structures that appear to serve different tasks.[8]

MAPPING NEURAL CIRCUITS

Tracing the human brain's wiring pattern is a daunting task. Decades ago, techniques that involved "sacrificing" the subjects, precluding human studies, were used to map neural circuits in animals. Attempts to develop postmortem mapping of circuits in humans were unsuccessful.[9] However, the continuity of evolution that Charles Darwin stressed throughout *On the Origin of Species* suggested using tracer studies of monkey brains to gather information on human neural circuits. Chimpanzees are closer to humans and might have provided a better approximation, but chimpanzees are an endangered species. Hence monkeys were the subjects.[10]

About twenty years ago, I was about to drive to a neuroscience workshop in Vermont when I looked into the mirror while shaving. My face was all puffed up. The thought came into my mind that I looked like a caricature of Benito Mussolini, the fascist dictator of Italy in World War II. What had happened was that the herpes type 1 virus that had been lurking in my body for decades had traveled through nerve pathways into my lips and surrounding facial tissue. I was advised simply to wait a few days for the swelling to subside.

Traditional tracer techniques make use of the tendency of some viruses to attach themselves to the chemical/electrical processes by which information is transmitted between neurons, the basic computing elements of all neural circuits. Certain chemical agents also have this property, and different tracers can be used to tag the neurons of a circuit by "propagating"—going up or down the circuit. If the objective is to trace the neural circuits that control an animal's tongue movements, a "back-propagating" tracer tag can be injected into the tongue to trace the circuit backward toward the animal's brain. A "forward-propagating" tracer tag injected into a neural structure can determine whether it is part of the circuit that ultimately controls the movements of the tongue. What places conventional tracer techniques completely outside the range of human experimentation is what necessarily follows. The animal is allowed to live for a week or so to allow the tracer to propagate through the neural circuit, and then it must be "sacrificed"—killed. The excised brain tissue then is stained with color couplers similar to those used in photographic color film. The color couplers—vivid reds and blues—attach to the tracer tag that had propagated through the neural circuit. The stained brain tissue is sliced in an apparatus that produces exceedingly thin sections. The sectioned slices then are microscopically examined and photographed. The result is an image of the circuit highlighted in color. A complementary, almost equally invasive technique can directly monitor neural activity by driving microelectrodes into an animal's brain to pick up the electrical signals that transmit information between neurons.

Studies using multiple microelectrodes generally confirmed the pathways mapped out by tracers in monkeys.[11] Direct electrode recording sometimes can be used to record activity in humans, for example, when electrodes are inserted into a patient's brain to stimulate a neural structure that can mitigate tremor in Parkinson's disease or reduce the severity of epileptic seizures.

RECYCLING BRAINS

One of the most vexing problems in evolutionary biology is determining the behavioral significance of changes, particularly when they involve brains. For example, it's clear that over the course of millions of years the average size of the brains of hominins has increased, which clearly is significant in light of the energy "cost" involved in keeping a larger brain going. And the computer analogy noted in earlier chapters seems valid too—with increased storage space, tasks that were impossible become routine. But the evolution of modern human capabilities seems to involve more than increasing brain size. Has excessive attention been focused on comparatively small increases in the limited sample of extinct hominin skulls that have been unearthed? Did an extinct hominin whose skull is somewhat larger than another fossil hominin, and had a brain that was 10 percent larger, possess cognitive capabilities superior to the smaller-brained hominin? In present-day humans, a person having a brain that's 10 percent larger does not necessarily enjoy cognitive ability superior to someone with a smaller brain, even if they have the same body mass.

THE SHAKING PALSY AND
THE BASAL GANGLIA

Similar issues attend interpreting what specific neural circuits do. When a neural circuit is mapped out, it is necessary to understand what aspects of behavior it may regulate. Experiments in nature, such as the Mount Everest hypoxia study discussed in chapter 1, continue to play a critical role in addressing this question. The Mount Everest hypoxia study would never have been undertaken if my colleagues and I had not already studied the motor and cognitive deficits associated with Parkinson's disease. As chapter 2 noted, Dr. James Parkinson, in his 1817 *An Essay on the Shaking Palsy*, described the tremor that marks the disease. In the late 1990s, studies aimed at ameliorating some of the problems of Parkinson's disease through surgery identified neural circuits that linked regions of the cortex with the basal ganglia and other subcortical structures. The circuits regulate motor control, including that involved in talking, and a range of cognitive capabilities, including some involved in language.[12]

Refer again to figure 1.1, which shows the basal ganglia, a set of neural structures deep within the human brain. The basal ganglia were first noted in the Roman era, though the names of the structures that form the basal ganglia were coined in the seventeenth century. William Willis, a British physician, then correctly thought that the basal ganglia were involved in motor responses to environmental stimuli. The caudate nucleus seemed to anatomists to have a tail—*caudate* means "tail" in Latin. The putamen resembled a lens. Detailed anatomical maps of the human basal ganglia were published in the first two decades of the nineteenth century. The root cause of Parkinson's disease is still unknown, but it is linked to the deterioration of the substantia nigra, a subcortical structure closely tied to the basal ganglia. The substantia nigra (its color is dark, hence its Latin name *black*) produces dopamine. Different neurotransmitters are active in different parts of the brain. Dopamine is only one of the neurotransmitters involved in Parkinson's disease; the neurotransmitter GABA plays a role in the substantia nigra's dopamine production.

Dopamine depletion was identified in the 1960s as the "proximate," that is, immediate, cause of Parkinson's disease. Dopamine cannot pass through the blood-brain barrier, which selectively

separates substances in the bloodstream from the fluid that cir-
culates in the brain. Levodopa, a dopamine precursor, can pass
through the blood-brain barrier. Levodopa converts to dopamine
after it enters the patient's brain. Along with other drugs, it's used
to treat Parkinson's disease. However, in the years before levodopa
was available, surgery on the subcortical structures of the basal ganglia was the only possible treatment. The findings of tracer studies
on monkeys, such as those reviewed by Garrett Alexander and
his colleagues in 1986, revealed a complex bundle of parallel circuits
involving the basal ganglia that clinical studies of Parkinson's disease indicated were involved in motor control, cognition, regulating
emotion, and attention.

You wouldn't think that cutting out part of someone's brain
could help them, but it was the only treatment available before
levodopa. David Marsden and Jose Obeso, in their 1994 review of
the results of Parkinson's surgery, came to the conclusion that the
basal ganglia could be thought of as a neural "switching system"
that could pull out and start up an "instruction" stored in motor
cortex, then stop it and pull out another "instruction," to produce a
sequence of motor acts, "submovements" that together constitute a
voluntary motor act such as walking. Taking a step while walking
upright, for example, first involves swinging your leg forward, then
tilting your heel down, then rocking forward, and so on.

No rational engineer would have designed the basal ganglia
switching system, in which activity is inhibited or initiated in an
overly complex manner, but it explains, up to a point, the effects of
brain surgery that mitigated some of the problems of Parkinson's.
In Parkinson's, the degraded basal ganglia switching system doesn't
properly carry out the sequence of submovements that constitute
the internally directed act of walking. A common clinical observation is that Parkinson's patients who have extreme difficulty in
walking can walk better if they copy someone walking. When

Parkinson's patients follow an external guide, they can carry out manual tasks that they otherwise cannot. Marsden and Obeso proposed that the basal ganglia also acted as a cognitive "switch" that allowed a person to change the direction of a thought process as well as a motor response when circumstances dictated a change— that is, beyond motor programming, the basal ganglia is also implicated in cognitive flexibility. This accounted for the "subcortical dementia" associated with Parkinson's that had become apparent in the 1980s. As is the case for the hypoxic Everest climbers discussed earlier, Parkinson's patients perseverate—they have difficulties changing the direction of their thought processes.

Marsden and Obeso also noted paradoxes—effects that they could not explain and still cannot be explained if we actually do know all the details of the circuits and the local operations of the linked neural structures. I used to advise students to avoid reading the Marsden and Obeso paper close to bedtime. The exact location of the surgical lesions discussed in the paper was known because patient after patient died a few weeks after surgery. In place of surgery, electrodes inserted into the basal ganglia are now used to control tremors by blocking local neural activity through electrical stimulation. The technique is similar in intent to jamming a broadcast by transmitting an interfering signal on the same frequency.

The initial functions of the basal ganglia may have been to carry out motor acts necessary to sustain life. However, basal ganglia recycling appears to have started millions of years ago in animals as far removed from humans as the ancestors of present-day lizards. In controlled experiments, lizards demonstrate a surprising degree of flexibility in response to changing environments.[13] Independent studies on species ranging from frogs to humans show that the basal ganglia also play a role in associative learning.[14] The details of what the basal ganglia usually do in associative learning were discovered in experiments that used electrodes inserted into the brains of mice and other animals, again precluding similar studies in humans.

However, diffusion tensor imaging (DTI), a noninvasive technique based on magnetic resonance imaging, confirms that humans have circuits similar to those found in monkeys linking regions of the cortex with the basal ganglia.[15] DTI has also revealed a web of circuits in humans connecting practically every part of the brain—the problem is to determine what these circuits do.

THE CASE OF THE WOMAN WHO SUDDENLY SPOKE IRISH

In my teens, I read, and reread, the *Case Book of Sherlock Holmes*, which included some bizarre cases. Sherlock Holmes came to mind when my colleague of many years Sheila Blumstein said that I might be interested in something odd—a woman who suddenly began to speak Irish.

Sheila's field of study focuses on aphasia. A local hospital had consulted her about a puzzling case—a woman who when she recovered from a coma suddenly began speaking with a strong Irish accent. It seemed to be an instance of a mysterious aphasic condition, "foreign-accent syndrome." The patient, CM, was a forty-year-old woman who wasn't Irish. She had never been to Ireland, and as far as anyone knew she didn't even have any close Irish friends. She had collapsed and went into a coma after taking a newly prescribed medicine that interacted with medications she had been taking for another problem. In 1997, when these events took place, medical specialists treating a patient could not access their patients' complete medical history.

My graduate student Emily Pickett, who was about to start her Ph.D. thesis research, and I decided to investigate. What we heard when we carefully listened to CM wasn't an Irish brogue. Computer-implemented acoustic analysis instead showed that CM's speech was peculiarly distorted. She was unable to coordinate and sequence properly the motor acts involved in talking. She couldn't keep the air pressure in her lungs steady or in synchrony with the muscles that moved her tongue, lips, jaw, and velum (a flap of soft tissue in the mouth that closes off the nose), and the muscles that control laryngeal phonation. Her speech would suddenly become loud or would abruptly fall to a whisper and then again suddenly become loud. Her vocal cords were phonating when they shouldn't have, in sounds such as [s]. The speech therapists at the hospital must have been at their wits' end when they decided that she was speaking with an Irish accent because her speech was unlike any that they had ever heard.[16]

Despite her distorted speech, CM could be understood. Her more debilitating problems were cognitive. A test of sentence comprehension showed that she had difficulties with sentences that six-year-old children could readily understand. Her life was constrained because she couldn't plan ahead. Her refrigerator door had a detailed daily schedule that listed routine tasks throughout the week. She couldn't adjust to unanticipated events, however minor. Similar problems afflict Parkinson's patients. The most severe cognitive problem, which CM shared with Parkinson's patients, was perseveration—the inability to change the direction of a thought process.

In a study published in 1985, Kenneth Flowers and Colin Robertson devised and used a simple measure of cognitive flexibility—the "Odd-Man-Out" sorting task—to study Parkinson's patients. Healthy control subjects usually are puzzled why anyone would bother administering such a simple test. The person being tested is presented with a booklet that has two sets of ten cards. Each card has three images printed on it. The first card might have a large triangle, a small triangle, and a large circle. The subject is asked to identify the "odd" image. There are two criteria the subject can use to reach that decision. The subject can use either shape or size to decide which image is "odd." If the criterion selected by the subject is size, the small triangle will be selected. If the criterion is shape, the circle will be selected. The second card also has three images printed on it, an uppercase *E*, a lowercase *e*, and an uppercase *A*. The subject is asked to select the odd image, using the same criterion. After each trial, the subject is told whether his or her decision was correct. The subject starts with the first ten-card set, and after he or she comes to the tenth card, the subject is asked "to do the sort another way" for the next ten-card set. After the next ten trials, the subject is again asked to resort to sorting the same packet of cards "the other way." This entails returning to the criterion that she or he used to sort the first ten-card set, five minutes or so earlier. The procedure is repeated six or eight times.

Parkinson's patients often have 30 to 50 percent error rates when they have to shift the sort either way—from size to shape or shape to size. The woman who spoke Irish, CM, was unable even to execute the first "set shift." She froze when she was asked to "do the sort another way" and asked Emily and me to please tell her what to do. The basal ganglia, like many structures of the brain, have left- and right-hemisphere—bilateral—segments. MRIs showed that her coma had produced large bilateral lesions in CM's caudate nucleus and putamen—structures of the basal ganglia that receive inputs from both motor cortex and prefrontal cortex. Brain-imaging techniques that monitor activity in living subjects' brains have since confirmed the role of the basal ganglia in changing the direction of a thought or action.[17]

AVOIDING HIGH-TECH FUMBA

As chapter 2 pointed out, it is possible to monitor brain activity safely in human subjects by means of functional magnetic resonance imaging (fMRI). The biological basis of fMRI is that brains "burn" glucose. fMRI monitors oxygen depletion, the byproduct of combustion. As a part of the brain become more active, it burns more glucose. fMRI tracks the relative level of local brain activity by tracking the local level of oxygen left after burning brain fuel. As more glucose is burned, the oxygen level falls, thus reflecting increased brain activity. The depleted blood oxygen level, the "BOLD" response that the computer in the fMRI system measures, is the signature of local brain activity. Direct comparison with neural activity monitored with microelectrodes inserted in the brain of animals confirms that the BOLD signal is a valid measure of neural activity.[18]

The technology is new, but phrenology fumba can pervade a fMRI study. The fMRI system's magnet, computer, and associated equipment provided the veneer of science for one study that ostensibly located the brain's "center of religion." Activity increased in the anterior cingulate gyrus when subjects, identified as "committed Christians," read sentences about Jesus performing miracles. Independent studies going back many years show that the anterior cingulate is activated when anyone pays attention to anything. The anterior cingulate was identified as an "erotic organ" in a different FMRI study in which young males viewed pornography.

fMRI studies have advanced the state of knowledge concerning the brain, but the initial problem that must be addressed in all fMRI studies follows from the reason for using fMRI. The subject being studied is alive. In living subjects, the circuits that control respiration, blood circulation, etc., are active, and their activity will register in the fMRI computer output as well as any brain activity related to the task being studied. One technique that often is used to discern whether the neural activity that has been recorded has anything to do with the behavior under study is to subtract the signals recorded from a baseline activity from the experimental condition. In an experiment that's aimed at discerning the brain mechanism involved in understanding spoken words, the subjects might be asked to first listen to musical tones. The fMRI signal relevant to speech could perhaps be determined by subtracting the activity recorded when the subjects listened to the tones from the activity recorded when they listened to the words. However, some neural mechanisms may be involved in both listening to musical tones and speech, so after the subtraction their activity wouldn't be apparent. In fact, many of the neural mechanisms active in speech perception also are active when you listen to music, so you might be wiping out relevant neural activity.[19] One solution to the ROI ("region of interest") problem is instead to monitor subjects as they perform similar tasks at increasing levels of difficulty. If activity in particular neural structures and circuits increase with task difficulty, it might be task related.

Other problems can derive from limits and compromises concerning computer processing. The massive magnet structure that you are placed into if you have an MRI provides signals that must be processed by a computer to visualize the structure of the brain or the neural activity in the brain. Processing limits often result in an fMRI study focusing on a region of interest, a predetermined region of the brain, excluding activity elsewhere. Studies that use ROIs aimed at the presumed site of Broca's area thus don't show activity elsewhere as a subject performs a language task. When the results of "whole-brain" studies are compared with studies limited to predetermined ROIs, it often is apparent that the presumed neural "language organ" or "brain center" of, for example, fear or pleasure does not exist. Sonja Kotz's fMRI experiment at the Max Planck Institute in Germany showed that contrary to what you might expect, activity increases when subjects have to focus solely on the emotion transmitted by the melody of speech than when they are asked to pay attention to the words of a sentence. Both the left and right hemispheres of the brain, including ventrolateral and dorsolateral prefrontal cortex, are implicated in this task.

Hundreds of fMRI studies have overlooked another inherent limitation. The fMRI signals derived after computer processing are weak and noisy—extraneous signals intrude, making it difficult to interpret the signal. One standard signal-processing technique used to solve this problem entails recording and averaging the fMRI signals from several subjects performing the same task, so as to factor out the noise. But the brains of ten different people vary, so if you want to see the activity in a particular part of the cortex, such as BA 44, the traditional site of Broca's area, there is a problem. While MRIs can show the boundaries of structures like the caudate nucleus, which has a distinctive shape and doesn't merge with any other structure, MRIs cannot discern the differences in the shape of the neurons in the layers of the cortex that define the Brodmann areas of the frontal regions of the cortex. The usual attempt to solve this problem entails stretching and squeezing the MRI images of the brains of each subject until they all conform to a "standard" brain, often that of a single sixty-year-old woman examined in 1988.[20] But that procedure is imperfect; the only precise way to determine the distribution of neurons in the six layers of the cortex that define each Brodmann area would be to sacrifice the subjects, then slice up and examine their brains microscopically.[21] The size of each Brodmann area also can vary dramatically from person to person. The size of BA 44, for example, varied tenfold in ten subjects.[22] One solution is not to attempt a fine-grain localization to Brodmann areas in prefrontal cortex, and instead note activity in larger anatomically defined regions, such as dorsolateral, ventrolateral, or orbofrontal prefrontal cortex.

Since the early days of fMRI studies, it also has been evident that the same area of the cortex may be active in different kinds of tasks, some linguistic, some not.[23] Ventrolateral prefrontal cortex (toward the lower part of the front side of your brain), for example, is active when people are asked to add or subtract numbers, comprehend the meaning of sentences, sort colors, recall words—and probably most other tasks.

INSIGHTS FROM fMRI

Nonetheless, fMRI and other neuroimaging studies have provided insights on the functional architecture and evolution of the human brain. It is clear that the neural bases of complex aspects of behavior, such as walking, talking, or comprehending the meaning of a sentence, are not "modules" that map onto areas of the cortex, nor do such subcortical modules exist. It instead appears to be the case that neural circuits linking activity in different parts of the brain are the brain bases for the acts that animals carry out, and for the thought processes of humans—and assuredly for the thoughts of animals. Since animals can't talk, there is a tendency to dismiss their capacity for thought, but studies of animal cognition strongly suggest that animals do think and plan.

It is becoming clear that the human brain supports hundreds, perhaps thousands, of different neural circuits. Noninvasive diffusion tensor imaging has revealed circuits connecting neighboring regions of the cortex as well as areas circuits connecting far-flung areas. DTI studies also show circuits linking regions of the cortex with subcortical structures—virtually everything seems to be connected to everything else. Some doubts have been raised as to whether all these circuits actually exist, and better imaging and mapping techniques have been called for, but it is clear that the neural bases of even "simple" acts are instantiated in circuits linking "local" activity in different parts of the brain.

However, what different parts of the brain do, or, for that matter, what different neural circuits do, is a work in progress. For example, the fMRI studies of Oury Monchi and his colleagues, which were aimed at understanding the neural mechanisms and deficits involved in Parkinson's disease, demonstrate that the operations performed by a neural circuit are not necessarily domain specific.

We don't appear to have separate neural systems that are committed to dealing with cognitive tasks involving visual patterns or sound patterns or the meaning of words. The Monchi group's fMRI data reveal similar patterns of neural activity in the caudate nucleus and putamen (basal ganglia structures) as well as linked prefrontal cortical areas of the cortex when subjects are asked to change the criteria by which they sort objects according to their color, shape, or number—or sort words according to their meaning, sound, or rhyming patterns.[24] The Monchi group's fMRI studies and other neuroimaging studies show that both ventrolateral prefrontal cortex and dorsolateral prefrontal cortex—two different areas of prefrontal cortex—enter into tasks that involve shifting cognitive criteria, recalling words and numbers in short-term "working memory," arithmetic, inhibiting thoughts and actions, comprehending the meaning of a sentence, judging emotion conveyed by intonation (the melody of speech),[25] and more. The "local" operations may vary in different tasks, which have yet to be established. What's clear is that activity registered in a particular part of the cortex while subjects perform task X does not indicate that that cortical region constitutes the "Faculty of X." As yet, we simply don't have clear answers on what different parts of the cortex do, or on the number, connections, and functions carried out by the neural circuits of the human brain.

Brains, like successful generals, also throw additional resources into action when things become difficult. One of the very first fMRI studies in 1996 showed that subjects activate additional cortical areas in order to comprehend sentences as they become more complex. Both left and right hemispheres of the cortex were active.[26] fMRI studies consistently show that as task difficulty increases, additional brain resources are brought into play and that it takes longer to complete the task. Thus, subjects listening to or reading sentences take longer to comprehend sentences that have more complex syntax. Their speech also slows down when they are performing cognitive tasks that are more demanding, perhaps owing to some neural structures being shared across different circuits involved in motor control and cognition or perhaps in circuits that do both. For example, as mental arithmetic tasks become progressively harder, the level of activity in ventrolateral prefrontal cortex increases, and subjects talk more slowly.[27] When animals or people learn new motor tasks, large areas of frontal cortex are activated. When the task is fully "automatic," only "motor cortex"—cortical areas associated with motor control—is active, and responses are rapid. The process by which motor acts become automatic—directed without conscious thought—involves forming "matrisomes," instruction sets by which groups of muscles are activated to carry out a task, in the frontal regions of the brain that direct motor control.[28]

At the height of the era of "Behaviorist" psychology, B. F. Skinner at Harvard University had a mouse circus. He had trained mice to perform amazing acts such as jumping onto and off trapeze bars. Skinner's theory of "Operant Conditioning" described the process by which the mice were trained, but stripped of jargon, each mouse either received a reward (food) or punishment (electric shock) for repeatedly performing or not performing one step of the circus act on command (usually a light going on). Eventually the mouse might learn to perform the linked steps. If neurophysiologic studies of the brains of the trained mice had been carried out, matrisomes in motor cortex most likely would have been found. Research started in the 1970s shows that frontal and subcortical regions of the brain as well as motor cortex are active while you are engaged in learning a task. However, when the task becomes automatic, most of the neural structures that were brought into action to learn the task no longer become active. They constitute a sort of temporary help that facilitate learning the task, but once it is learned, they're no longer needed. At the point when an action becomes automatic, matrisomes, populations of neurons that code the entire action, are formed in motor cortex. They code the submovements, the motor acts necessary to carry out the task automatically. Brain-controlled prostheses offer the hope of someday restoring normal life to amputees by connecting matrisomes to computer-controlled mechanical fingers, wrists, ankles, feet, etc.

Miyamoto Musashi's 1645 lesson book for samurai in *The Book of the Five Rings* described how to learn to wield the Samurai sword, to deadly effect—immediate action without conscious thought. The goal was to be able instantly to execute complex sword thrusts that would bite into a target. Musashi's lesson plan also works for learning to walk, to talk, to play the piano, and the complex protocols that mark human culture, including learning the linguistic "rules" of syntax. If you've taught someone how to drive a car with a stick shift, you've probably followed Musashi's formula. To master the gearshift and clutch, you must slowly perform the action, repeating it again and again. At first you'll have to think consciously about each step. But at some point you may realize that you are not aware of the steps involved and can go on to think about other things.

EVO-DEVO, TRANSCRIPTIONAL, AND EPIGENETIC FACTORS

Charles Darwin, in *On the Origin of Species*, followed the line of inquiry proposed by his grandfather Erasmus. The similarities that mark related species in their embryonic stage suggested the continuity of evolution. However, in the course of ontogenetic development to their adult forms, bodies morph. It is becoming apparent that transcriptional and epigenetic factors govern the course of ontogenetic development to form the human body. Ongoing studies suggest that in the course of evolution, transcriptional and epigenetic factors also contributed to the formations of the neural structures and circuits that confer the motor and cognitive capacities that distinguish humans from other living species. The discovery that the neural "machinery" involved in talking and thinking, and language includes circuits linking cortex with subcortical structures opened a new mystery. The circuits are very similar in monkeys and humans. Moreover, some of the linked neural structures in these neural circuits exist in frogs and lizards and seem to date back to the earliest transitional reptilian mammals. So why are other living species unable to talk, fully command language, or possess human creative ability?

Researchers have uncovered some hints about the evolution of the human brain. Research that started in the 1990s suggests that a series of mutations on transcriptional factors and epigenetic factors specific to humans in effect "supercharged" hominin brains. An "experiment in nature" and subsequent research using cutting-edge genetic technology identified one of the transcriptional factors and has pointed to further studies on the genes and epigenetic information involved. There clearly is more to the story of the evolution of the brain bases of human behavior than these events. As the previous chapter pointed out, upright posture, the ability to run for long periods of time, being able to exploit new sources of food, cooking, and increased capacities for social interaction and the ability to inhibit violent acts all appear to have played a role in crafting hominin bodies and brains. However, our abilities to think the oddest things and actions, share our thoughts by talking to one another, regulate emotional responses if we so wish, and spend our time in "useless" artistic activity seem to hinge on events that crafted the human brain in the last half million years.

A genetic anomaly discovered in an extended family in London and a debate on the validity of Noam Chomsky's claim that all humans possess an innate organ of the brain, a universal grammar (UG) that specifies the syntax of every language that has or will exist, triggered research that has addressed this question about the recent evolution of the human brain. Noam Chomsky, one of the founders of MIT's Department of Linguistics, where I earned my Ph.D. in the 1960s, proposed that all human beings are born with a universal grammar that allowed them to acquire language. Chomsky's focus was on the syntax of language, which he claimed is preloaded into every human brain. Children supposedly do not have to learn the particular syntactic "rules" of their native language. A child merely had to be exposed to a particular language to activate the particular preloaded innate "principles" and "parameters" that specified that language's syntax.

The speech, language, and cognitive deficits in some members of an extended family, identified as the KE family, in London had come to the attention of the Institute of Child Health of University College, London. Fifteen children and adults in the KE family had severe difficulty talking. On further study, the Institute of Child Health discovered that, additionally, an odd syndrome—a suite of behavioral problems—marked the afflicted members of the KE family. They had profound difficulties coordinating orofacial gestures and could not simultaneously stick out their tongues while pursing their lips. They could not repeat two words in sequence. They had problems choosing the correct words when attempting to speak to another person and had difficulties comprehending distinctions in meaning conveyed by simple English syntax. Motor control was awkward, and cognitive abilities such as "verbal working memory," remembering a sequence of words, was impaired. On standardized Wechsler intelligence tests, affected KE members had significantly lower scores than their unaffected siblings, a result that ruled out environmental factors influencing the results. The problem appeared to be genetic, deriving from one grandmother and transmitted autosomally to both males and females.[29]

In 1990, a report by Myrna Gopnik in the prestigious scientific journal *Nature* claimed that the language deficit of the afflicted members of the KE family was limited to a fragment of English grammar, the "regular" past tense and plural forms of verbs (e.g., *talk* versus *talked* and *cat* versus *cats*), while preserving other aspects of language, except, oddly, talking. Gopnik, a linguist who supported Chomsky's views, while visiting London had limited her tests of the syntactic capabilities of the KE family to this aspect of English syntax, which is used in every introductory class on Noam Chomsky's linguistic theory to demonstrate that English syntax is governed by a small set of innately specified rules of grammar. The example figures into many textbooks and into books, such as Steven Pinker's, aimed at the general public. However, what's omitted is that when more examples of English syntax are examined, it becomes evident that thousands of different rules are necessary. Pinker publicized Gopnik's findings, which were reported in *Nature*, and he organized a special session that focused on them at the annual meeting of the American Association for the Advancement of Science. If the report in *Nature* was correct, the singular language problem in the KE family would support Chomsky's view that the details of grammar were genetically built into "normal" brains and merely had to be activated. An isolated genetic anomaly in the afflicted members of the KE family had destroyed a fragment of the UG—the rules of syntax that governed regular English nouns and verbs.

In the course of the debate that followed when members of the Institute of Child Health asked *Nature* to publish their own findings, which refuted Gopnik's claims, I was sent a videotape of a BBC documentary that focused on the KE family. The BBC had filmed five of the affected children who were in a remedial program, their teachers, and one of the Institute of Child Health's geneticists. Profound speech deficits were evident. One child, a five-year-old girl, had just begun to be able to say her name. The BBC documentary used English subtitles because the child's speech was so distorted as to be incomprehensible. It was impossible to understand what she was saying even when she concentrated on pronouncing the first syllable of her name. The other children had somewhat greater success, but their speech had the mechanical quality of a poorly designed speech synthesizer. The children's teachers were upbeat, but they pointed out that the children also had difficulties beyond those of mastering the regular forms of past-tense verbs and plurals. The children had general problems choosing the right words and syntax when they attempted to produce any sentence. They had difficulties understanding all aspects of English syntax, including sentences that conveyed distinctions using irregular verbs and nouns. The documentary confirmed the findings of the paper that the Institute of Child Health had submitted to *Nature*, which should have ended the claim that the KE family's problems with language proved that Noam Chomsky's universal grammar really exists, but still today you can find references to the KE data to support the argument that UG is part of the human brain's innate "Faculty of Language."

The Institute of Child Health researcher who had appeared in the BBC documentary also held to the Broca's area fumba, stating that the KE family's problems derived from damage to Broca's area. About a year later, I was in London at the Institute of Child Health because the daughter of a close Sherpa friend in Nepal seemed to be having similar problems. She was born in Lo Manthang, a village that then lacked any medical facilities. She had difficulty talking, and when she was brought to the United States to assess her condition, a MRI revealed damage to her brain that had disrupted the neural circuits linking prefrontal and posterior regions of the cortex with the basal ganglia as well as to the hippocampus, cerebellum, and other subcortical structures.

The profile of her behavioral problems included some of the symptoms seen in Parkinson's disease and in strokes that damage the basal ganglia or the neural circuits that link it to the cortex and other parts of the brain. At the Institute of Child Health, my assessment of the neural basis of the behavioral problems of the KE family members was that they seemed to be similar to those observed when circuits linking the cortex and the basal ganglia were disrupted. A few years later, the ICH reported the results of an MRI study that revealed anomalies in the afflicted KE family members' basal ganglia. Simon Fisher, a geneticist at Oxford University, then found that the affected members of the KE family had only one copy of a transcriptional factor, the *FOXP2* transcriptional gene, in place of the normal two copies.[30] In the human brain, *FOXP2* acts on the development of the neural structures that form the human cortical-striatal-cortical circuits involved in motor control and cognition; it also acts on the development of the cerebellum, which is involved in cognition.[31] The cerebellum is an "ancient" part of the brain, a part shared with all mammals; it traditionally was thought to be concerned only with motor control. Cerebellar disease results in poor motor coordination and timing, but it also results in cognitive and linguistic deficits. Neuroanatomical studies showed that *FOXP2* also acts on the embryonic development of the lungs, intestinal system, heart, and other muscles in both mice and humans.

This pointed to a new avenue in exploring the human brain's evolution. When the genetic assay from the newly available chimpanzee genome project was compared with human DNA, it was evident that the human version of *FOXP2* differed from the chimpanzee version. A "human" version, *FOXP2^human^*, had evolved during the six-to-seven-million-year period that separates the evolution of humans and chimpanzees from their common ancestor. During that period, *FOXP2^human^* underwent two substitutions in its DNA sequence from chimpanzee *FOXP2^chimp^*. Two of the amino acids in its DNA code had changed—breakneck speed compared to the sole amino-acid substitution in the 70 million years of evolution between mice and chimpanzees.[32]

A knock-in technique, in which a gene or genes from a different species is inserted into a genome, was used by the genetics unit of the Max Planck Institute for Evolutionary Biology in Leipzig to establish the probable effects of *FOXP2* on hominin brains. The human form *FOXP2^human^* replaced the "wild," natural form of *Foxp2* (the lowercase letters signify that it is not the human form) in mouse pups. When *FOXP2^human^* was knocked into mouse pups, their vocalizations were somewhat different than the normal calls of mouse pups that retained wild *Foxp2*. When the wild form of *Foxp2* was knocked out in mouse pups, they died soon after birth, which was not surprising in light of *Foxp2*'s role in muscle, lung, and cardiovascular development.[33] However, the evolutionary significance of the mutations that yielded *FOXP2^human^* was not the vocalizations produced by the mice.

When the cortical–basal ganglia neural circuits of the mice that had received the *FOXP2^human^* gene were studied, synaptic plasticity was enhanced in basal ganglia neurons. Dendrite lengths also increased in the basal ganglia, thalamus, and layer VI of the cortex. Dendrites are the connections that branch out from neuronal cell bodies to transmit information to other neurons. *FOXP2^human^* mice also had increased synaptic plasticity in the connections between neurons of the basal ganglia and the substantia nigra, which forms part of the complex basal ganglia circuits and produces the neurotransmitter dopamine.

The process by which we learn anything—motor acts, names, concepts, etc.—involves modifying synaptic "weights," that is, the degree to which synapses transmit information between neurons. Synaptic weights undoubtedly changed in the neural motor-control circuits in the samurai who followed Miyamoto Musashi's training program. Donald Hebb in 1949 first pointed out the role of synaptic modification as the basis by which animals learn and store information. Synaptic modification is the neural mechanism by which the relations that hold between seemingly unrelated phenomena are learned. In animals such as mollusks, which don't have a central brain, synapses in their tails that control motion are modified when the mollusks learned to associate the smell of squids (a mollusk delicacy) with an electric shock. In mice, increasing synaptic plasticity in their basal ganglia circuits enhances associative motor learning. The mouse knock-in study also showed that $FOXP2^{human}$ affected the cortical plate (layer 6), the input level of the cortex.

TIME TRAVEL—BRAINS AND VOCAL TRACTS

Though we cannot witness events that occurred five hundred thousand, two hundred thousand, or even one hundred years ago, genetic time travel is now a reality. The Max Planck Institute genetics research group directed by Svante Paabo pioneered techniques that allow us to recover and analyze the DNA of long-dead hominins. The form of $FOXP2^{human}$ bearing two amino-acid substitutions from the chimpanzee version is shared by every living human and by Neanderthals.[34] That places its appearance to between 370,000 and 450,000 years ago, when humans and Neanderthals last had a common ancestor. The time estimates are from a genetic "clock" that takes into account the amount of genetic divergence owing to the random mutations that constantly occur. *FOXP2^human* also was found in Denisovans, who may be a distinct species that split off from Neanderthals about 200,000 years ago.[35] Only a few fragments of Denisovan bones have been recovered from a finger in Siberia, but a large group of hominin fossils in a cave in Spain may be early Neanderthals or Denisovans, or perhaps hominins that were ancestral to both. Further study seems to be warranted.[36]

Humans also have human-specific epigenetic information on the DNA strand to close to where *FOXP2* is coded (intron 9), and a selective sweep appears to have occurred about 200,000 years ago that may have enhanced human cognitive capabilities. The study of the genetic and neural bases of speech and language has progressed far from the traditional Broca-Wernicke theory. Genevieve Konopka and Todd Roberts in their 2016 review article on the neural and genetic bases of vocal communication in the journal *Cell* note other human species–specific genetic and epigenetic modification on the neural circuits that link different regions of the human cortex with the basal ganglia and cerebellum. Some epigenetic factors also appear to have enhanced cortical activity. Neuroimaging studies suggest that almost all areas of the cortex are implicated in linguistic tasks.[37] Genes, transcription factors, and epigenetic information, some specific to humans, have acted on the human cortex and the elements of circuits linking the cortex with subcortical structures. And the neural bases of language and speech are not completely separable from those involved in regulating motor control, cognition, emotion, and even basic sensory modalities such as vision.

Moreover, it is now apparent that a massive epigenetic restructuring of the genes that determine the anatomy that enables us to talk and sing occurred during the same time period—after anatomically modern humans (AMH) split from Neanderthals and Denisovans. David Gokhman, at the Hebrew University of

Jerusalem, and his colleagues have provided new insights on the processes that underlie the differentiation of skeletal and soft tissues. DNA methylation is one of the primary indicators of the epigenetic mutations that affect the expression of genes—the processes that ultimately result in skeletal and soft tissue. Gokhman and his colleagues first reconstructed the methylated regions of a Neanderthal and a Denisovan fossil and compared them with a present-day human osteoblast (bone remodeling) methylation map.[38] From this map they established a comprehensive assembly of skeletal DNA methylation maps, adding a 40,000-year-old Neanderthal, four ancient anatomically modern humans who lived 7,000 to 40,000 years ago, and more than fifty-five present-day humans. This comparison enabled them to identify differentially methylated regions (DMR) of the human genome. A comparison with six chimpanzee methylomes established the "derived" hominin and the variations in their sample of modern humans. They were able to filter out within-population variations, arriving at a set of 6,371 DMR between the human and Neanderthal-Denisovan groups. They then used a procedure that links genes to organs to identify organs that are significantly overrepresented in DMRs, and found that genes that affect speech and voice quality were most enriched in modern humans. The tissues most affected were those of the larynx and pharynx. After an exhaustive analysis, they concluded that "voice-affecting genes are the most over represented DMGs along the AMH lineage, regardless of inter-skeletal variability, coverage by methylation array probes, the extent to which a DMR is variable across human or chimpanzee populations, or the significance levels of DMRs."

In short, after accounting for genetic variability and taking into consideration known mutations that affect voice quality and facial structure, Gokhman and his colleagues note that it was evident "that genes affecting vocal and facial anatomy have gone through particularly extensive regulatory changes in recent AMH evolution." They conclude that

> these regulatory changes played a key role in the shaping of the human face, as well as in the formation of the unique 1:1 configuration of the human vocal tract that is optimal for speech. Our results provide insight into the molecular mechanisms that formed the modern human face and vocal tract, and suggest that they arose after the split from the Neanderthal and the Denisovan.

Edmund Crelin and I could never have imagined in 1971, when we proposed that Neanderthal vocal anatomy could not produce the full range of human speech, that the genetic bases for the evolution of the human vocal tract would ever be established. Species-specific human epigenetic mutations enhanced the robustness of human speech. In about the same timeframe, the brain mechanisms that allow us to learn and execute the motor acts involved in speech were enhanced by mutations on the *FOXP2* transcriptional gene and by epigenetic factors that again appear to be specific to humans. No ape has been able to talk, and the motor acts that are involved in talking are among the most complex activities that "ordinary" people attain. Although a full understanding of the neural bases of human cognitive ability and language is still a work in progress, it is clear that the human neural circuits that regulate motor control also are involved in aspects of cognition and language that set us apart from other living species: planning, decision-making and reasoning, complex syntax, verbal working memory, and the creative capacities evident in virtually all aspects of human behavior.

These aspects of human behavior—our ability to talk, thereby sharing our moods, desires, needs, plans, and the store of information that we each learn during our lifetime and the aggregated knowledge of a culture or civilization—lead to success in "the struggle for existence." As Charles Darwin would have pointed out, as the complexity of culture and the knowledge base of human cultures increased, so did the selective value of mutations that augmented the capacities of vocal communication and language. The inseparable relation that holds between the environment and biology is stressed throughout *On the Origin of Species.*

IN SHORT

The traditional Broca-Wernicke theory for the neural bases of language is wrong. Paul Broca in 1860 resuscitated phrenology, the early-nineteenth-century pseudoscience that located "faculties"— neural "organs" that each regulated a specific aspect of human behavior—in specific bumps and hollows of a person's skull. Broca thought that the aphasic language deficits of the two patients whose brains he examined after they died from strokes derived from damage to a specific region of the cortex—Broca's area—but it is apparent that aphasia, the permanent loss of language, results from damage to subcortical structures of the brain that support neural circuits. That, however, does not mean that the human cortex is not involved in language. Cortical areas appear to be malleable; for example, areas that form parts of circuits involved in vision can take on auditory tasks in blind people. Neural circuits that link local activity in different parts of the brain, both cortical and subcortical areas, regulate complex aspects of behavior such as walking, talking, and language. Studies of behavioral deficits after strokes, other instances of brain damage, and neurodegenerative disease point to the presence of these neural circuits.

Tracer studies that can map out neural circuits in animals and noninvasive techniques that can monitor brain activity in human subjects confirm the role of these neural circuits. Neural circuits linking regions of the cortex with the basal ganglia and other sub-cortical neural structures play a role in linguistic comprehension, cognitive acts such as changing the direction of an action or thought, or holding information in short-term "working memory." The architecture of these circuits in monkeys and humans is similar, raising the question of why humans can talk, command complex language, and possess a degree of cognitive flexibility greater than any other living species.

As Darwin observed in *On the Origin of Species*, the body forms that distinguish species occur as living organisms transition from their embryological to their adult state. "Evo-Devo" studies show that transcriptional and epigenetic factors govern the process of transcription and ultimately an animal's anatomy and brains. Ongoing studies suggest that transcriptional and epigenetic factors appear to have acted on hominin brains at an accelerated rate during the past 500,000 years to, in effect, supercharge the human brain. The transcriptional factor $FOXP2^{human}$, which differs from its chimpanzee version, produces enhanced synaptic plasticity and connectivity in human and Neanderthal brains. Epigenetic factors specific to humans appear to have played a role in shaping the human brain. Present-day humans whose ancestors left Africa in the migrations that populated Eurasia and Australia have a small number of genes inherited from Neanderthals or Denisovans, reflecting mating between humans and these extinct species, and an ensemble of epigenetic events specific to humans clearly shaped the species-specific human vocal tract and larynx as well as the shape of the human head. Both the brain bases of human language and thought and the anatomy that makes human speech a more robust form of communication appear to have reached their present stage after modern humans diverged from Neanderthals and Denisovans.

5

WHAT WOULD DARWIN
THINK ABOUT . . .

We don't have a time machine that could bring Charles Darwin into the twenty-first century, but *On the Origin of Species* and other writings suggest how he would have viewed current issues of general concern. I'll start with the easy questions.

GLOBAL WARMING

In light of Charles Darwin's observation that a seemingly minor human action, such as building a fence, can trigger a chain of events that creates a very different ecosystem, it is clear that he would be alarmed by climate change brought about by recklessly burning fossil fuels. The data would be indisputable—Greenland's glaciers melting, torrential rain flooding cities, rising sea levels. Darwin would be a presence on the Internet, the twenty-first century's penny post, blogging, tweeting, and posting on Facebook. And he would be e-mailing friends and contacts in high places, urging them to action.

POPULATION GROWTH

Malthus's essay on the chaos that could have resulted from unchecked population growth triggered Charles Darwin's "eureka" moment. The disaster predicted by Malthus hasn't happened (yet), but it was apparent to Darwin that population growth posed a problem. In chapter 3 of *On the Origin of Species*, which discusses this issue, the first example is: "Even slow-breeding man has doubled in twenty-five years, and at this rate, in a few thousand years, there would literally not be standing room for his progeny."[1]

Subsequent events bear out this prediction. One of the strongest impressions I've retained from several trips to India is that of crowds—masses of people. In 1959, India's population was 359 million; by 2015, it had reached 1.25 billion. In parts of the world it is now impossible to grow enough food to feed everyone. Water will soon cost more than oil and may not even be available. Charles Darwin would clearly be using his standing to urge measures for population control that go beyond the "prudent restraint from marriage" noted in *On the Origin of Species*.

ECOLOGY AND EXTINCTION

Chapter 3 also serves as a tract on how humans affect ecosystems. The effects of human intervention on the balance of nature were apparent in the world of country estates and sleepy villages in which Charles Darwin moved. Darwin pointed out the changes he witnessed as open heaths were fenced and put to use:

> Twelve species of plants (not counting grasses and carices [grassy plants and herb]) flourished in the plantations, which could not be found on the heath. The effect on insects must have been still greater,

for six insectivorous birds were very common in the plantations, which could not be seen on the heath: and the heath was frequented by two or three distinct insectivorous birds. Here we see how potent has been the introduction of a single tree, nothing whatsoever else having been done, with the exception that the land had been enclosed, so that cattle could not enter.[2]

Britain's colonial empire also was expanding. *Simmonds's Colonial Magazine and Foreign Miscellany*, for example, which argued for exploiting the colonies to free England from relying on foreign commodities, justified its position thus in an article printed in 1845: "Will any man be bold enough to assert that these fair portions of the earth have been created by an all-wise and munificent Providence for no other purpose than to be the haunts of wild beasts?"

Darwin's views on the "primitives" from whom these wild lands would have to be seized were formed early in his life. His comments in *The Voyage of the* Beagle concerning the Fuegian passengers reveal that he viewed all humans as members of the same species, all sharing the same foibles. He formed a strong belief that culture determined the state of civilization—whether you lived almost unclothed, pelted by sleet, in Tierra del Fuego, or attended the opera in evening dress. A tainted version of "the struggle for existence" was used by Haeckel and others to justify the superiority of the "white race" and colonization, but Charles Darwin undertook no active part in the formation of the British Empire. Nor did he view as "improvements" such policies as enclosing the commons that had marked English villages for a thousand years.

Extinction is an inevitable consequence of evolution. *On the Origin of Species* is explicit: "for as new forms are continually and slowly being produced . . . numbers inevitably must become extinct."[3] Almost every morning, the e-mails in my inbox include calls for urgent action to save an endangered species. I'm asked to contact my senator or member of Congress and also click on a link to send money. Apart from not knowing whether most of the contribution will be used to pay for e-mails and the organization's staff, it sometimes is the case that for the species in question, extinction is inevitable.

One trigger for extinction is specialization to a narrow environmental window. The argument for maintaining genetic diversity cuts two ways. If a species is adapted to an extraordinarily narrow niche, then change disrupting that niche may result in extinction, unless extraordinary measures are taken.

In 1975, the Tennessee Valley Authority had almost completed the construction of the Tellico dam on the Little Tennessee River. The project had cost 80 million dollars and would have provided clean power to an economically depressed region. The opponents of constructing the dam seized on the Endangered Species Act, which had been passed a few years earlier, and insisted that construction had to be halted because it would wipe out the snail darter, a tiny perch whose total population had been estimated at only a few hundred. The fish were thought to live only in the Little Tennessee River, and the dam would block their migration route. The courts agreed, but the endangered status of the snail darter was lifted in 1978 after they were transplanted to other rivers and discovered elsewhere. The dam was completed, and no one seems to have had much further interest in snail darters.

In this instance, everything worked out for everyone, but would a better use of the money for the darter-recovery plan have been to further early-childhood education in Tennessee? In 2015, Tennessee was still ranked at C– and was twenty-eighth out of fifty states on the Early Education Index. Charles Darwin might have thought that in this instance the Endangered Species Act had been misused.

However, in other instances, forceful action is warranted to prevent slaughter motivated by greed feeding on stupidity. In the 1870s, a fashion craze began for large-brimmed women's plumed hats adorned with bird feathers and even small stuffed animals. By 1886, four years after Charles Darwin's death, five million birds were being killed each year. The snowy egret was almost driven to extinction in Florida by plume hunters. The Royal Society for the Protection of Birds and the Audubon Society in the United States joined together to stop plume hunting. In 1901, Florida passed a bird-protection act; Teddy Roosevelt established the first wildlife sanctuary in 1903. But bird feathers were worth their weight in gold, so poaching continued.

Plumed hats eventually went out of fashion, and bird populations recovered. Elephant tusks and rhinoceros horns now are in high demand, but the governments that should protect these species are ineffective and often corrupt. In Darwin's *Descent of Man*, which included material that could have been published in 1859, Darwin made his views on conservation clear when he wrote: "Disinterested love for all living creatures, the most noble attribute of man."[4]

The plumed-hat craze and the concomitant slaughter of birds did not reach its apogee until after his death, but it is clear that Charles Darwin would have been forcefully acting to end that slaughter; were he alive today, he would be protesting the poaching of Africa's animals to extinction.

GMOs

Grafting, a form of genetic engineering, has been practiced for thousands of years. It was in use in China at least four thousand years ago, and from there spread through Asia and Europe. The goal in grafting is to produce a genetically identical copy of a plant or tree that has desirable properties or a hybrid that has superior properties. The upper part of a plant, the "scion," is inserted into the rootstock of another plant. Virtually all of the apples that you eat come from grafted trees. Peaches, pears, oranges, and nuts almost always come from trees that have been grafted. Camellias, the flowers associated with the American South, are usually produced by grafting, as are roses and other flowers. Superior watermelons can be produced by grafting on squash rootstock. Budding is a similar, more recent technique, in which the buds of a plant are transferred.

Charles Darwin might be puzzled at the shape of the present controversy concerning genetically modified organisms, GMOs. To many people, genetically modified food is anathema—"Frankenfood" developed by greedy corporations and presenting a danger to health and the environment. GMOs are created by inserting a gene or genes from one species into the DNA sequence of a different species, or by deleting a gene. The technique can be viewed as "focused grafting" that specifies the specific gene that you wish to transfer from one plant to another. It is also possible to modify animals genetically, though it is more difficult, and the outcomes may not be viable.

GloFish, genetically modified zebrafish in bright blue, green, yellow, and other fluorescent colors, were produced by the same technique used to knock in the $FOXP2^{human}$ transcriptional factor in mice. Genes from jellyfish, coral, and other organisms were inserted into zebrafish embryos, allowing them to merge with the zebrafish's genome. GloFish can be purchased online in the United States. They have been smuggled into the Netherlands, since the European Union bans GMOs. Though GloFish are not supposed to be eaten and were never approved for human consumption, fluorescent sushi recipes are on the Internet. The U.S. Food and Drug Administration did approve one fish—genetically modified salmon—for human consumption in 2015.

The GMO debate seems to involve issues that transcend the techniques used to create GMOs and their safety. In the United States, it is focused on the Monsanto Corporation, a multinational biotechnology company, and Monsanto's "Roundup-Ready" seeds. Monsanto manufactures "Roundup," a herbicide used for weed control; the active ingredient is glyphosate, which acts on an enzyme found in plants. Genetically modified corn and soybean seeds resistant to Roundup were developed, enabling farmers to spray entire fields with Roundup without also killing their corn and soybeans. The GMO seeds are patented, and Monsanto aggressively sues farmers for license fees for crops harvested from GMO seeds that happened to drift onto their land. The lawsuits are perceived as mean-spirited and unfair. New studies also are raising questions on the safety of long-term exposure to low levels of glyphosate, which may increase the risk of cancer.

However, Monsanto and other biotechnology companies are producing GMO seeds that don't involve using Roundup. These seeds are resistant to insects and drought. One newer GMO product increases the yield of soybeans. Similar seeds for other crops in a food-deprived world will be useful. "Golden" rice, a GMO that was developed by university scientists and biotechnology companies, can alleviate vitamin A deficiencies in millions of Asian children. In an open letter published on June 30, 2016, in the *New York Times*, 107 Nobel laureates criticized Greenpeace's stand opposing any GMOs. The letter included references to studies and review papers spanning years that found GMO crops to be safe and not actually different from other crops. In the Nobel laureates' opinion, GMOs also offer the best chance of being able to feed the planet's rapidly growing population.

There is always an element of risk when any new species is introduced into an ecosystem. The U.S. Department of Agriculture has a website listing invasive plants and animals—"natural" ones, not GMOs—that drive out native species and in some cases have caused irreparable environmental damage. In light of the record of problems caused by invasive species that Darwin observed, he clearly would have advised caution when "releasing" any GMO, but it is also clear that he would not be lining up with Greenpeace or other organizations to oppose all GMOs. Virtually everything we eat has been modified through human intervention, starting with the wild berries, apples, potatoes, and grains that our distant ancestors first cultivated.

DARWIN DENIERS

The *New York Times*, on June 20, 2016, reported that a 450-foot-long wood structure, costing more than $102 million dollars, was scheduled to open near the Creation Museum in Williamstown, Kentucky.[5] It's supposed to be a replica of Noah's Ark.

The ark, which may be the largest timber-frame building in the world, is intended to serve as a warning that, according to the Bible, God sent a flood in Noah's time to wipe out a depraved people and that God will deliver a fiery end to those who reject the Bible and accept modern-day evils like abortion, atheism, and same-sex marriage. The ark's builder, Ken Ham, founded the Creation Museum to further the beliefs of the Young Earth Creationism movement, which started about seventy years ago. According to Young Earthers, the Hebrew Bible is a historical record: the universe is six thousand years old, and Darwin's theory of evolution is wrong and leading to our moral and mortal downfall. The movement's ministry has published books, magazines, videos, and curriculums used by thousands of churches and homeschoolers.

In 1844, Christians who did not accept the literal interpretation of the Bible that drives the Young Earthers could take comfort in Robert Chambers's book *Vestiges of the Natural History of Creation*. The earth has changed, species that once lived are now extinct, and new species, related to ancestral species, have evolved. However, these events are seen as part of a master plan ordained by God. That premise, rebranded as "intelligent design," is the position followed today if you don't necessarily take the stories in Genesis and other creation myths as facts but still believe in an intelligent entity that governs the universe. This theoretical entity can account for phenomena observed by evolutionary biologists such as transcriptional genes, epigenetic promoters, and so on. They all would be part of a plan beyond human comprehension that is gradually coming to the attention of science. Surprisingly, some academics on the left of the political spectrum, such as Noam Chomsky, are among the most vociferous exponents of intelligent design.

THE INTELLIGENT DESIGN
OF LANGUAGE

Noam Chomsky has, over the course of fifty years, elaborated a theory on the evolution of human language close to that of Robert Chambers. As hypothesized in *Vestiges*, the neural basis of human language is an innate "Faculty of Language," an organ of the brain that is species specific and regulates language and language alone— phrenology reborn. And Chomsky is a Darwin denier. Since the 1970s, he has argued that natural selection has a negligible role, if any, in the evolution of living organisms. In his 1972 book *Language and Mind*, Chomsky states: "It is perfectly safe to attribute this development [of language] to 'natural selection,' so long as we realize that there is no substance to this assertion, that it amounts to no more than a belief that there is some natural explanation for these phenomena."[6]

The subtext of Chomsky's *The Science of Language: Interviews with James McGilvray* is a sustained attack on natural selection. Upbraiding the evolutionary biologist Richard Dawkins for misleading people about natural selection, Chomsky writes: "Tell them the truth about evolution, which is that selection plays some kind of role, but you don't know how much until you know. It could be small, it could be large; it could even be nonexistent."[7]

Comments that stress the absence of evidence for natural selection pervade *The Science of Language*. Summing up Chomsky's views on language, McGilvray, echoing Chomsky, states that "you [Chomsky] emphasize in the case of human language—certainly the most distinctive and central mental faculty, one that no other creature has . . . there is no evidence that a long-term selectional story will work. There are reasons to believe that language was introduced at a single stroke with the introduction of Merge, perhaps some fifty or sixty thousand years ago." "Merge" is the capacity for "taking two things and putting them together or taking one thing and taking a piece of it and sticking it at the edge."[8] Merge accounts for what Chomsky takes to be the defining feature of human language—being able to construct and comprehend complex sentences that include clauses or conjunctions. In *The Science of Language* and other articles and books, Chomsky claims that no form of language existed earlier than 50,000, 80,000, or 100,000 years ago (the date keeps shifting from one publication to the next), when Merge suddenly came into every human brain.[9]

However, how did Merge come into the brain of every human 50,000, 60,000, 80,000, or 100,000 years ago—the exact date is immaterial? Evolutionary biologists can point to events such as the selective sweep in which the genes that confer adult lactose tolerance was transmitted by means of natural selection acting over time to most of the individuals who lived in cultures where herds of milk-producing animals were present. But natural selection does not exist in Chomskyland: he has declared that the concept of natural selection has "no substance." If challenged, Chomsky probably would have no difficulty accounting for how Merge spread through the humans who left Africa. The course of evolution to Chomsky instead is intelligent design, a process directed by some unspecified force or agent that somehow transformed everyone's brains. It's possible, though improbable, that some epigenetic factors were involved, in similar fashion to the ones that reduced the lifespans of the grandchildren of the women who lived in nineteenth-century Sweden's far north. Or perhaps, as the Young Earthers and other Creationist sects would hold, Merge proves that God exists and governs the world—Merge is a miracle.

A DARWINIAN MODEL FOR
THE EVOLUTION OF LANGUAGE

Evidence concerning the nature and evolution of the biological bases of human language wasn't present in Darwin's time, nor were studies that have since demonstrated affinities between human language and the latent and realized capabilities of other species. These data support a Darwinian model for the evolution of human language—a gradual process over a long period of time.

Darwin, in *On the Origin of Species*, pointed out the recycled human tongue, which plays a role in the evolution of human language: "The strange fact that every particle of food and drink which we swallow has to pass over the orifice of the trachea, with some risk of falling into the lungs."[10] The ontogenetic development of the human supralaryngeal vocal tract (SVT), which yields the strange, low position of the human larynx, was discussed in chapter 2. Independent computer-modeling studies of the acoustic consequences of maneuvers of the tongue, lips, and other structures that can change the shape of the SVT show that an adultlike human SVT is necessary to produce the "quantal" vowels specific to human speech. Quantal speech sounds, such as the vowel [i] (the vowel in the word *see*), are less susceptible to errors in tongue placement. Their acoustic characteristics also make them less susceptible to being misidentified. Moreover, the vowel [i] provides an optimal signal in the process that yields the high data-transmission rate of human speech. Chapter 2 briefly reviewed the physiology and acoustics of quantal speech sounds.

The soft tissue of the supralaryngeal vocal tract (SVT) is not apparent in the fossil record, but in 1971, Edmund Crelin and I reconstructed the SVT of the Neanderthal fossil skull that had been unearthed early in the twentieth century near the village of La Chappelle aux Saints in France. The Neanderthal skull shared affinities with the shape of newborn human skulls and appeared to have had a similar tongue and SVT. Computer modeling of the reconstructed Neanderthal SVT showed that it could not have produced quantal [i] sounds.

Neanderthal cervical vertebrae were subsequently unearthed that established Neanderthal neck lengths. Robert McCarthy, who with Daniel Lieberman had previously shown that the neck length and vocal tracts of children do not reach adult proportions until the age of six to eight years,[11] and I used this information to refine the reconstruction of the SVTs of Neanderthals and other archaic hominins.[12] Neanderthal SVTs had similar mouth-neck proportions as five-year-old children, which computer modeling showed yielded almost, but not quite, quantal speech capabilities. In contrast, Australopithecine SVTs would have had the same speech-production limits as apes. The SVT reconstructions showed that the human SVT had gradually evolved. It was also apparent that the neural circuits that allowed humans to learn and execute the complex motor acts that underlie speech must have been in place before the human SVT evolved. There otherwise would not have been any selective advantage in retaining a tongue-larynx configuration that increased the risk of death from choking when eating.

The basal ganglia and other neural structures that perform local operations in these neural circuits have long and deep evolutionary histories. The basal ganglia, as the previous chapter noted, date back to early amphibians.[13] The intonation of speech, which entails controlling the larynx, also involves neural structures that have a long evolutionary history. Intonation, the controlled modulation of laryngeal phonation, signals sentence boundaries and other aspects of syntax. Controlled fundamental frequency contours, the product of laryngeal phonation, also differentiate words in tonal languages such as Mandarin Chinese.[14] The neural structures and circuits that control the fundamental frequency of phonation can be traced back to therapsids, mammal-like reptiles who lived during the Triassic, Jurassic, and early Cretaceous eras. These neural structures may have initially evolved to facilitate mother-infant communication. Studies of human mother-infant communication reveal a special vocal mode or "register"—the "motherese" dialect by which parents address infants.[15] Motherese speech has a high fundamental frequency of phonation and extreme pitch variations. Mammalian infants (including human infants) also signal for attention by means of a forceful isolation cry that has a high pitch and amplitude—the cry that can immediately rouse parents from a deep sleep.

Comparative studies suggest that therapsids possessed the anatomical specializations and neural structures to produce isolation cries. All mammals possess a paleocortex, which includes the anterior cingulate cortex (ACC). Studies of the effects of damage to the ACC and the neural circuits that connect it to other parts of the brain show that the anterior cingulate cortex plays a role in controlling F0 and pitch; the ACC also directs attention to virtually anything that one wishes to do. While the soft tissue of the brains of therapsids has not survived, the inference that these mammal-like reptiles had an ACC rests on the fact that they possess three middle-ear bones, which are found in all present-day mammals. The initial function of the hinge bones of the reptilian jaw was to open the jaw wide. In the course of evolution, the hinge bones were recycled to function as "organs" of hearing. The transition from mammal-like reptiles to mammals involved the former jaw bones migrating into the middle ear to enhance auditory acuity.

All mammals have both an ACC and these middle-ear bones, so middle-ear bones are regarded as an index for the presence of the ACC. Vocal communication enhancing mother-infant contact clearly contributes to reproductive success. The neural circuits involving the ACC appear to have been recycled to facilitate both attention and laryngeal control. Lesion studies on mice, for example, show that mouse mothers do not pay attention to their infants when neural circuits to the ACC are disrupted. Virtually every neuroimaging study ever published shows ACC activity when subjects are asked to perform any task, both directed and involuntary.

WHO HAD WHAT SORT OF LANGUAGE?

In light of the absence of innovation apparent in the archaeological record—the stasis evident in Acheulian hand axes—and the comparatively recent 500,000-years-ago date for the genetic and epigenetic events that shaped the hominin brain, it is unlikely that *Homo erectus* and earlier hominins possessed full human cognitive capability. Claims that *Homo erectus* possessed full-blown language because they may have crossed the Pacific on rafts[16] are best regarded as improbable as Chomsky's claim for no form of language existing before 100,000 or 60,000 years ago.

However, in light of the capabilities of chimpanzees and the archeological record, *Homo erectus* probably possessed some aspects of language. *Homo erectus* left evidence of their presence in Europe and Asia and of their ability to survive in hostile environments, suggesting linguistic capabilities superior to those apparent in chimpanzees. It is impossible to determine the form of any *Homo erectus* language, but words probably were spoken—sounds that in themselves had no meaning but to the members of a community signified objects and actions. Perhaps the linguistic process started earlier, in Australopithecines who, through association, noticed that they could determine that some fruits were ripe when they turned a certain color and communicated this information.

Orangutans can copy intonation patterns,[17] and early hominins surely had the same capability, so a F0 pattern that signified the color red perhaps could have become a shared vocal symbol signifying ripe fruit. It is impossible to test this speculation, but it is probable that language evolved gradually. As is the case for other unique aspects of human behavior, such as walking and being able to stand erect, the course of evolution involved variation, natural selection, and, perhaps, direct heritable environmental effects over millions of years.

NATURE VERSUS NURTURE

Evolutionary psychologists such as Steven Pinker and Marc Hauser have claimed that other aspects of human behavior have an innate basis. As I pointed out earlier, Hauser proposed an innate, genetically transmitted basis for moral conduct similar in nature to Chomsky's universal grammar.[18] Steven Pinker claimed that everyone has a preference for images or drawings that show a few trees and a pool of water on an almost flat plain.[19] This art "instinct" supposedly reflects preferences imprinted in every human brain when our distant ancestors lived on the savannahs of Africa millions of years ago. One difficulty that Pinker doesn't address is that many extinct hominins lived in heavily forested regions of Africa, but the fallacy is that no data support his claim for everyone preferring that scene. As noted earlier, some studies claim that infants are born with an innate "core" knowledge of physics.

Charles Darwin instead stressed the interplay of the environment and the evolution of innate capacities. The selective advantage conferred by a genetic variation is always subject to environmental constraints. Genes that could confer adult lactose tolerance would have little value absent herds of milk-producing animals. A mere fence can trigger a chain of events that changes the environment. Closing off the common land on which animals could graze resulted in natural selection favoring different species of plants, insects, and birds. Darwin summed up the synergy between the environment and natural selection when he stated that "any variation, however slight and from whatever cause proceeding, if it be in any degree profitable to an individual of any species, in its infinitely complex relations to other organic beings and to external nature, will tend to the preservation of that individual, and will generally be inherited by its offspring."[20]

THE GOD QUESTION

Though his Unitarian grandfather and father were agnostics, Charles Darwin was inculcated in high-church Anglican doctrine in boarding school and at Cambridge University. The short autobiography that he began to write in 1876 for his children and grandchildren reveals his journey toward uncertainty and disbelief. The complete autobiography was published in 1958 by his granddaughter Norah Barlow, who restored the deletions made after Darwin's death. In his autobiography, Darwin wrote:

> Whilst on board the *Beagle* I was quite orthodox, and I remember being heartily laughed at by several of the officers (though themselves orthodox) for quoting the Bible as an unanswerable authority on some point of morality. I suppose it was the novelty of the argument

that amused them. But I had gradually come, by this time, to see that the Old Testament from its manifestly false history of the world, with the Tower of Babel, the rainbow as a sign, etc., etc., and from its attributing to God the feelings of a revengeful tyrant, was no more to be trusted than the sacred books of the Hindoos, or the beliefs of any barbarian.[21]

Darwin continues by noting his hesitancy to relinquish his faith.

But I was very unwilling to give up my belief; I feel sure of this for I can well remember often and often inventing day-dreams of old letters between distinguished Romans and manuscripts being discovered at Pompeii or elsewhere which confirmed in the most striking manner all that was written in the Gospels. But I found it more and more difficult, with free scope given to my imagination, to invent evidence which would suffice to convince me. Thus disbelief crept over me at a very slow rate, but was at last complete. . . . I can indeed hardly see how anyone ought to wish Christianity to be true; for if so the plain language of the text seems to show that the men who do not believe, and this would include my Father, Brother and almost all my best friends, will be everlastingly punished.

And this is a damnable doctrine.[22]

The autobiography was written in bits and pieces. Later, Darwin charts the inner mental voyage that led to his concluding that God is an invention—a belief created by the human mind.

Although I did not think much about the existence of a personal God until a considerably later period of my life, I will here give the vague conclusions to which I have been driven. The old argument of design in nature, as given by Paley, which formerly seemed to me so conclusive, fails, now that the law of natural selection has been discovered.

We can no longer argue that, for instance, the beautiful hinge of a bivalve shell must have been made by an intelligent being, like the hinge of a door by man. There seems to be no more design in the variability of organic beings and in the action of natural selection, than in the course which the wind blows. Everything in nature is the result of fixed laws.[23]

Would Darwin be an outspoken atheist publicly if he were alive? Probably not. His wife Emma believed that she would reunite in Heaven with the sister who died early in life and with her dead children. She hoped that Charles's disbelief would not bar him from reuniting with her in eternal life. Although Emma's religious convictions lessened as she and he grew old, Charles took care to avoid offending her.

Like the Buddha, Darwin would continue to avoid answering the God question.

NOTES

1. STRAWBERRIES

1. Charles Darwin, *On the Origin of Species* (1859; Cambridge, Mass.: Harvard University Press, 1964), 41–42.
2. Darwin, *On the Origin of Species*, 45.
3. Darwin, *On the Origin of Species*, 30.
4. Darwin, *On the Origin of Species*, 34, 12, 27.
5. Darwin, *On the Origin of Species*, 36.
6. J. Browne, *Charles Darwin, Voyaging* (Princeton, N.J.: Princeton University Press, 1995), 245.
7. Charles Darwin, *The Voyage of the* Beagle (1839; London: John Murray, 1905), 207–208.
8. Darwin, *The Voyage of the* Beagle, 205, 213.
9. Darwin, *The Voyage of the* Beagle, 229.
10. In Browne, *Charles Darwin, Voyaging*, 339.
11. Darwin, *On the Origin of Species*, 52.
12. Darwin, *On the Origin of Species*, 61.
13. As the rainfall changes, so does the proportion of nuts that are easier to crack. Natural selection acts on the finches; when their diet mostly consists of hard nuts, the finches tend to have shorter, thicker beaks; soft nuts result in longer, thinner beaks. If Darwin had spent about thirty years in the Galapagos, he might have observed these events.
14. S. Rasmussen, M. E. Allentoft, K. Nielson, et al., "Early Divergent Strains of *Yersinia pestis* in Eurasia 5,000 Years Ago," *Cell* 163 (2015): 571–582.

15. A. L. Kolata, *Ancient Inca* (New York: Cambridge University Press, 2013), 263, 26.

16. S. A. Tishkoff, F. Reed, et al. "Convergent Adaptation of Human Lactase Persistence in Africa and Europe," *Nature Genetics* 39 (2007): 31–40. In different places, at different times, adult lactose tolerance spread throughout populations through selective sweeps on different genes.

17. M. Fumagalli, L. Moltkee, N. Garup, et al., "Greenland Inuit Show Genetic Signatures of Diet and Climate Adaptation," *Science* 349 (2015): 1343–1347.

18. In the nineteenth century, Spanish mining engineers in the Andes discovered that virtually everyone's physical abilities deteriorate when they spend more than two weeks above 5,300 meters. Respiratory physiologists have known for decades that almost no one can adapt to altitudes that exceed 5,300 meters; their performance on mental and physical tasks slowly deteriorates. A. Bouhuys, *Breathing* (New York: Grune and Stratton, 1974).

19. T. S. Simonson, Y. Yang, C. D. Huff, et al., "Genetic Evidence for High-Altitude Adaptation in Tibet," *Science* 329 (2010): 71–74.

20. P. Lieberman, B. G. Kanki, et al., "Cognitive Defects at Altitude," *Nature* 372 (1994): 325; P. Lieberman, A. Morey, et al., "Mount Everest: A Space-Analog for Speech Monitoring of Cognitive Deficits and Stress," *Aviation, Space, and Environmental Medicine* 76 (2005): 198–207.

21. B. L. Smith, "Temporal Aspects of English Speech Production: A Developmental Perspective," *Journal of Phonetics* 6 (1978): 37–68, found that children speak slowly and don't talk as fast as most adults until age ten. Carol Chomsky, *The Acquisition of Syntax in Children from Five to Ten* (Cambridge, Mass.: MIT Press, 1969), found that children did not master the syntax of their native language until age ten. Noam Chomsky, who maintains that language is "acquired" in the first years of life, could not have overlooked his wife's book.

22. P. Lieberman, "On the Nature and Evolution of the Neural Bases of Human Language," *Yearbook of Physical Anthropology* 45 (2002): 36–62; P. Lieberman, *Toward an Evolutionary Biology of Language* (Cambridge, Mass.: Harvard University Press, 2006); P. Lieberman, *The Unpredictable Species: What Makes Humans Unique* (Princeton, N.J.: Princeton University Press, 2013).

23. K. A. Flowers and C. Robertson, "The Effects of Parkinson's Disease on the Ability to Maintain a Mental Set," *Journal of Neurology, Neurosurgery, and Psychiatry* 48 (1985): 517–529.

24. E. R. Pickett, E. Kuniholm, A. Protopapas, et al., "Selective Speech Motor, Syntax, and Cognitive Deficits Associated with Bilateral Damage to the Putamen and the Head of the Caudate Nucleus: A Case Study," *Neuropsychology* 36 (1998): 173–188.

25. O. Marin, W. J. Smeets, and A. Gonzalez, "Evolution of the Basal Ganglia in Tetrapods: A New Perspective Based on Recent Studies in Amphibians," *Trends in Neurosciences* 21 (1998): 487–494.

26. Darwin, *On the Origin of Species*, 72.

2. NO CATS, NO FLOWERS

1. W. Stukeley, *Memoirs of Sir Isaac Newton's Life* (1752), http://www .royalsociety.org/turning-the-page.

2. Charles Darwin, *The Autobiography of Charles Darwin, 1809–1882*, restored ed., ed. Nora Barlow (London: Collins, 1958), 120.

3. Charles Darwin, *On the Origin of Species* (1859; Cambridge, Mass.: Harvard University Press, 1964), 63.

4. S. Pinker, *How the Mind Works* (New York: Norton, 1998); S. Pinker, *The Language Instinct: How the Mind Creates Language* (New York: William Morrow, 1994). The immediate source of Pinker's model of the brain is Jerry Fodor's *Modularity of Mind* (Cambridge, Mass.: MIT Press, 1983). Fodor and Pinker claim that brains have the same "modular" design as conventional electronic devices and digital computers: that is, each particular part controls one particular function. The radars used in World War II had circuit boards that each controlled a particular function—the visual display, the range-estimating circuits, the power supply, etc. The modular design enabled operators who were not expert electrical engineers to repair a unit rapidly by exchanging "modular" circuit boards. But biological brains don't follow modular design principles. Fodor himself, in some of his talks and publications, points out that his theory is best regarded as neophrenology.

5. Darwin, *On the Origin of Species*, 62–63.

6. Darwin, *On the Origin of Species*, 73, 77.

7. Charles Darwin, *Notebook E., Transmutation (1838–1839)*, 63. http://darwin -online.org.uk/EditorialIntroductions/vanWyhe_notebooks.html.

8. J. Goodall, *The Chimpanzees of Gombe: Patterns of Behavior* (Cambridge, Mass.: Harvard University Press, 1986), 503.

9. W. Paley, *Natural Theology; or, Evidences of the Existence and Attributes of the Deity* (London: R. Faulder, 1802).

10. Darwin, *On the Origin of Species*, 108–109, 194.

11. Darwin, *On the Origin of Species*, 190.

12. Chu-Yu Tseng's 1981 computer-implemented analysis of three different Chinese languages ("An Acoustic Study of Tones in Mandarin," PhD diss., Brown University, 1981) shows the pitch patterns that differentiate words in different Chinese languages when words are carefully enunciated. In normal conversations the pitch patterns are often absent—listeners "hear" them by a subjective process in which the context indicates the appropriate word. Similar phenomena occur in all languages—we often hear what we think should be in the acoustic signal. When we know what someone's talking about, we hear more detail.

13. L. Lisker and A. S. Abramson, "A Cross-Language Study of Voicing in Initial Stops: Acoustical Measurements," *Word* 20 (1964): 384–442.

14. Arend Bouhys's *Breathing* (New York: Grune and Stratton, 1974) explains respiratory physiology in a jargon-free manner. It's clear that massive recycling and chance events have created a weird system that works but that is hardly the work of an intelligent planner. For details on speech production, see P. Lieberman, *Toward an Evolutionary Biology of Language* (Cambridge, Mass.: Harvard University Press, 2006); P. Lieberman, *The Unpredictable Species: What Makes Humans Unique* (Princeton, N.J.: Princeton University Press, 2013).

15. Darwin, *On the Origin of Species*, 191; http://www.nsc.org/library/report_injury_usa.htm.

16. Edmund S. Crelin noted the apelike position of the larynx in his *Anatomy of the Newborn: An Atlas* (Philadelphia: Lea and Febiger, 1969). The complex developmental process leading to the adult SVT was charted by Daniel Lieberman and Robert McCarthy in "The Ontogeny of Cranial Base Angulation in Humans and Chimpanzees and Its Implications for Reconstructing Pharyngeal Dimensions," *Journal of Human Evolution* 36 (1999): 487–517. They measured cephalometric X-rays taken in the 1940s that tracked the development of the skull, neck, and tongue from birth to age twenty. The details are discussed in D. E. Lieberman, *The Story of the Human Body: Evolution, Health, and Disease* (New York: Pantheon, 2013); and P. Lieberman and R. C. McCarthy, "The Evolution of Speech and Language," in *Handbook of Paleoanthropology*, 2nd ed., ed. W. Henke and I. Tattersall (Berlin: Springer, 2015).

17. P. Lieberman, D. H. Klatt, and W. H. Wilson, "Vocal Tract Limitations on the Vowel Repertoires of Rhesus Monkey and Other Nonhuman

Primates," *Science* 164 (1969): 1185–1187. Fitch and colleagues, guided by cineradiographic films of a rhesus monkey eating and vocalizing, modeled the potential vowel space, arriving at a vowel repertoire similar to modeling studies reported in the 1970s for newborn humans and chimpanzees. The monkey vowel repertoire reported in this study again did not include the vowel [i] or other fully quantal vowels. W. T. Fitch et al., "Monkey Vocal Tracts Are Speech Ready," *Science Advances* 2 (2016): e1600723. Fitch and de Boer in this paper falsely characterize the findings of their own studies, claiming that anatomy played no part in the evolution of language. Their own papers actually state the opposite—that the unique human vocal tract enhances the communicative value of human speech; see P. Lieberman, "Comment on 'Monkey Vocal Tracts Are Speech Ready'; Monkey Business: Did the Evolution of Speech Involve Anatomy?" *Science Advances* 3 (2017): e1700442.

18. G. E. Peterson and H. L. Barney, "Control Methods Used in a Study of the Vowels," *Journal of the Acoustical Society of America* 24 (1952): 175–184.

19. J. Kaminski, J. Call, and J. Fisher, "Word Learning in a Domestic Dog: Evidence for Fast Mapping," *Science* 240 (2004): 1676–1671.

20. E. S. Spelke, K. Breinlager, J. Macomber, et al., "Origins of Knowledge," *Psychological Review* 99 (1992): 605. Spelke and likeminded psychologists who support Noam Chomsky's nativist theories claim that infants possess "core-knowledge" of physics, moral conduct, and arithmetic. The basis for their claims are experiments in which infants view edited films where they see a ball rising instead of falling when someone drops it, to which the infants evince surprise. Recent studies show that human infants, like other creatures, pay more attention to unexpected events. See A. Stahl and E. Feigenson, "Observing the Unexpected Enhances Infants' Learning and Exploration," *Science* 348 (2015): 91–97. The infants, by the time that they view Spelke's films, have seen many objects falling down. Only a few things they have seen, such as helium-filled balloons at birthday parties, rise.

21. The elegant techniques used in these experiments are described in J. P. Gavornik and M. F. Bear, "Learned Spatiotemporal Sequence Recognition and Prediction in Primary Visual Cortex," *Nature Neuroscience* 17 (2014): 732–737; and F. Sam, S. F. Cooke, R. W. Komorowski, et al., "Visual Recognition Memory, Manifested as Long-Term Habituation, Requires Synaptic Plasticity in V1," *Nature Neuroscience* 18 (2015): 262–273.

22. J. J. Shea, *Stone Tools in Human Evolution: Behavioral Differences Among Technological Primates* (New York: Cambridge University Press, 2017).

23. Isabel Gauthier and her colleagues, using the same fMRI technique as Kanwisher, showed that the FFA is active when car and bird experts view pictures of cars and birds. I. Gauthier, P. Skudlarski, J. C. Gore, and A. W. Anderson, "Expertise for Cars and Birds Recruits Brain Areas Involved in Face Recognition," *Nature Neuroscience* 3 (2000): 191–197. Gauthier and Tarr showed that after some practice at identifying strange-looking objects, "greebles," the FFA becomes active when people look at greebles. I. Gauthier and M. J. Tarr, "Unraveling Mechanisms for Expert Object Recognition: Bridging Brain Activity and Behavior," *Journal of Experimental Psychology: Human Perception and Performance* 28 (2002): 431–446. The FFA seems to be part of a visual system that facilitates recognizing objects, scenes, or animals that are of interest to an individual.

3. GRANDFATHER ERASMUS

1. Erasmus Darwin, *Zoonomia; or, The Laws of Organic Life*, vol. 1., section 39, "On Generation," part 8, http://books.google.co.uk/books?id=A0gSAAA AYAAAJ. Erasmus Darwin's "filament" proposal is from ibid., part 7.

2. J. Browne, *Charles Darwin: The Power of Place* (Princeton, N.J.: Princeton University Press, 2002), 450. The often-repeated claim that Darwin "stole" natural selection from Alfred Russel Wallace is incorrect. Darwin's notebooks show that his "eureka" moment occurred in 1838 when he read Malthus's essay on the disaster that the increase in population would produce. Wallace, moreover, viewed natural selection as a regulatory mechanism that acts to keep a species adapted to the ecosystem in which it lives. In the 1858 paper read at the Linnean Society of London along with extracts from Darwin's 1844 sketch of his theory and an 1857 letter to Asa Gray, Wallace stated that "the action of this principle is exactly like that of the centrifugal governor of the steam engine, which checks and corrects any irregularities almost before they become evident; and in like manner no unbalanced deficiency in the animal kingdom can ever reach any conspicuous magnitude, because it would make itself felt at the very first step, by rendering existence difficult and extinction almost sure soon to follow."

3. F. Darwin, ed., *The Life and Letters of Charles Darwin, Including an Autobiographical Chapter* (London: John Murray, 1888), 149–150.

4. E. Mayr, introduction to *On the Origin of Species*, by Charles Darwin, facsimile ed. (Cambridge, Mass.: Harvard University Press, 1964), xxii.

5. C. Darwin, *On the Origin of Species* (1859; Cambridge, Mass.: Harvard University Press, 1964), 20.

6. Darwin, *On the Origin of Species*, 22–23.

7. Darwin, *On the Origin of Species*, 485.

8. F. Darwin, ed., *The Life and Letters of Charles Darwin*, 14.

9. Darwin, *On the Origin of Species*, 138.

10. B. Flammang et al., "Tetrapod-like Pelvic Girdle in a Walking Cavefish," *Scientific Reports* (2016). The waterfall-climbing cave fish *Cryptotora thamicola* has evolved many of the skeletal features of our hominin ancestors, including a full-blown pelvis that facilitates walking. They provide evidence of "parallel" evolution: natural selection directed at achieving a similar end in different species.

11. Darwin, *On the Origin of Species*, 85.

12. A. Forsdahl, "Are Poor Living Conditions in Childhood and Adolescence an Important Risk Factor for Arteriosclerotic Heart Disease?" *British Journal of Preventative and Social Medicine* 31 (1977): 91–95.

13. M. E. Pembrey et al., "Sex-Specific, Male-Line Transgenerational Responses in Humans," *European Journal of Human Genetics* 14 (2006): 159–166. M. E. Pembrey et al., "Human Transgenerational Responses to Early-Life Experience: Potential Impact on Development, Health, and Biomedical Research," *Medical Genetics* 9 (2014): 563–572.

14. S. K. Reilly et al., "Evolutionary Changes in Promotor and Enhancer Activity During Human Corticogenesis," *Science* 347 (2015): 1155–1159. For a comprehensible review of epigenetics, see D. Dominissini, C. He, and G. Rechavi, "RNA Epigenetics," *The Scientist* (2016), http://www.the-scientist.com/?articles.view/articleNo/44873/title/RNA-Epigenetics/.

15. J. Kagan, J. Resnick, and N. Snidman, "Biological Bases of Childhood Shyness," *Science* 240 (1988): 167–171.

16. Studies of heritable epigenetic effects can be accessed almost daily in journals such as *Cell*, *Science*, *Nature*, *PNAS-USA*, and online journals such as *E-Life*.

17. Darwin, *On the Origin of Species*, 484.

18. Darwin, *On the Origin of Species*, 482.

19. Darwin, *On the Origin of Species*, 479.

20. Darwin, *On the Origin of Species*, 485.

21. Over the course of thirty years, Peter and Rosemary Grant showed that natural selection rapidly adapted the beaks of the Galapagos finches to

breaking open the nuts that they ate. Shifts in rainfall brought on by the El Nino Pacific Ocean current changed the ratio of hard to soft nuts. Natural selection resulted in the finches' beak size and shape rapidly adapting to changes in the food supply. When hard nuts were abundant, beaks became shorter and stouter. When soft nuts were abundant, beaks became thinner. In light of the Grants' study and scores of classic and recent studies, it's difficult to understand the basis for Noam Chomsky's claim that no evidence for natural selection exists. The Grants' thirty-year-long study was described by Jonathan Weiner in his Pulitzer Prize–winning book *The Beak of the Finch: A Story of Evolution in Our Time* (New York: Knopf, 1994).

22. Browne, *Charles Darwin*, 325–329.

23. J. Goodall, *The Chimpanzees of Gombe: Patterns of Behavior* (Cambridge, Mass.: Harvard University Press, 1986), 590.

24. V. Reynolds et al., "Mineral Acquisition from Clay by Budongo Forest Chimpanzees," *PLoS One* (July 28, 2015), http://dx.doi.org/10.1371/journal .pone.0134075.

25. S. J. Harmand et al., "3.3-Million-Year-Old Stone Tools from Lomekwi 3, West Turkana, Kenya," *Nature* 521 (2015): 310–315.

26. S. Savage-Rumbaugh, D. Rumbaugh, and K. McDonald, "Language Learning in Two Species of Apes," *Neuroscience and Biobehavioral Reviews* 9 (1985): 653–665; R. A. Gardner and B. T. Gardner., "Teaching Sign Language to a Chimpanzee," *Science* 165 (1969): 664–672.

27. J. Shea, *Stone Tools in Human Evolution: Behavioral Differences Among Technological Primates* (New York: Cambridge University Press, 2017).

28. R. Corbey et al., "The Acheulean Handaxe: More Like a Bird's Song Than a Beatles Tune?" *Evolutionary Anthropology: Issues, News, and Reviews* 25 (2016): 6–19.

29. J. Zilhãoa et al., "Symbolic Use of Marine Shells and Mineral Pigments by Iberian Neanderthals," *Proceedings of the National Academy of Sciences* 107 (2010): 1023–1028.

30. C. Henshilwood et al., "A 100,000-Year-Old Ochre Processing Workshop at the Blombos Cave, South Africa," *Science* 334 (2011): 219–221.

31. S. McBrearty and A. Brooks., "The Revolution That Wasn't: A New Interpretation of the Origin of Modern Human Behavior," *Journal of Human Evolution* 39 (2000): 453–563.

4. CRAFTING THE HUMAN BRAIN

1. For example, in R. C. Berwick et al., "Evolution, Brain, and the Nature of Language," *Trends in Cognitive Sciences* 17 (2013): 89–98.

2. C. Wernicke, "The Aphasic Symptom Complex: A Psychological Study on a Neurological Basis [1894]," in *Proceedings of the Boston Colloquium for the Philosophy of Science*, vol. 4, ed. R. S. Cohen and M. W. Wartofsky (Dordrecht: Reidel, 1967).

3. K. Brodmann, "Beiträge zur histologischen Lokalisation der Grosshirnrinde. VII. Mitteilung: Die cytoarchitektonische Cortexgleiderung der Halbaffen (Lemuriden)," *Journal für Psychologie und Neurologie* 10 (1908): 287–334; K. Brodmann, "Ergebnisse uber die vergleichende histologische Lokalisation der Grosshirnrinde mit besonderer Berucksichtigung des Stirnhirns," *Anatomischer Anzeiger*, suppl., 41 (1912): 157–216; K. Brodmann, *VergleichendeLokalisationslehre der Groshirnrinde in iheren Prinzipien dargestellt auf Grund des Zellenbaues* (Leipzig: Barth, 1909).

4. N. Dronkers et al., "Paul Broca's Historic Cases: High-Resolution MR Imaging of the Brains of Leborgne and Lelong," *Brain* 130 (2007): 1432–1441.

5. P. Marie, *Traveaux et mémoires* (Paris: Masson, 1926).

6. D. T. Stuss and D. F. Benson, prominent specialists in aphasia, in their *The Frontal Lobes* (New York: Raven, 1986), directed at neurologists and specialists treating aphasia, pointed out that permanent loss of speech and/or language occurs only when damage to subcortical structures and pathways of the brain occurs. Patients who have suffered damage limited to the cortex, including Broca's and Wernicke's areas, generally recover within six months.

7. K. Goldstein, *Language and Language Disturbances: Aphasic Symptom Complexes and Their Significance for Medicine and Theory of Language* (New York: Grune, 1948).

8. For reviews of neural-circuit operations, see P. Lieberman, *Human Language and Our Reptilian Brain: The Subcortical Bases of Speech, Syntax, and Thought* (Cambridge, Mass.: Harvard University Press, 2000); P. Lieberman, "On the Nature and Evolution of the Neural Bases of Human Language," *Yearbook of Physical Anthropology* 45 (2002): 36–62; P. Lieberman, *Toward an Evolutionary Biology of Language* (Cambridge, Mass.: Harvard University Press, 2006); P. Lieberman, *The Unpredictable Species: What Makes Humans Unique* (Princeton, N.J.: Princeton University Press, 2013).

9. Kuypers thought that human neural circuits could be traced after death, but they cannot. He believed that he had discovered a species-specific human neural circuit that provided a direct channel from the cortex to the larynx, which he mistakenly believed was the "organ" of speech. H. Kuypers, "Corticobulbar Connections to the Pons and Lower Brainstem in Man," *Brain* 81 (1958): 364–388. But as noted here and elsewhere, the larynx is only one of the anatomical structures involved in talking. Moreover, the neural circuit that is implicated in human laryngeal control, as is the case in all other mammals, involves the basal ganglia. One of the symptoms of Parkinson's disease, in which the basal ganglia deteriorate, is loss of laryngeal control—see the discussion of this issue in P. Lieberman, "Vocal Tract Anatomy and the Neural Bases of Talking," *Journal of Phonetics* 40 (2012): 608–622; and in Lieberman, *The Unpredictable Species*.

10. Much of our knowledge of neural circuits comes from studies of the brains of rhesus macaque monkeys; see, for example, G. E. Alexander et al., "Parallel Organization of Segregated Circuits Linking Basal Ganglia and Cortex," *Annual Review of Neuroscience* 9 (1986): 357–381. For a review, see M. Petrides, "Lateral Prefrontal Cortex: Architectonic and Functional Organization," *Philosophical Transactions of the Royal Society B* 360 (2005): 781–795.

11. E. K. Miller and M. A. Wilson, "All My Circuits: Using Multiple Electrodes to Understand Functioning Neural Networks," *Neuron* 60 (2008): 483–488.

12. J. L. Cummings, "Frontal-Subcortical Circuits and Human Behavior," *Archives of Neurology* 50 (1993): 873–880; C. D. Marsden and J. A. Obeso, "The Functions of the Basal Ganglia and the Paradox of Sterotaxic Surgery in Parkinson's Disease," *Brain* 117 (1994): 877–897. The background material is reviewed in Lieberman, *The Unpredictable Species*.

13. Leal found that lizards adapt to changing environments, albeit slowly. M. Leal and B. J. Powell, "Behavioural Flexibility and Problem Solving in a Tropical Lizard," *Biology Letters* 8 (2012): 28–30.

14. For studies of the specific role of the basal ganglia in monkeys learning to perform tasks, see A. M. Graybiel, "The Basal Ganglia and Cognitive Pattern Generators," *Schizophrenia Bulletin* 23 (1997): 459–469; A. M. Graybiel, "Building Action Repertoires: Memory and Learning Functions of

the Basal Ganglia," *Current Opinion in Neurobiology* 5 (1995): 733–741; and J. Mirenowicz and W. Schultz, "Preferential Activation of Midbrain Dopamine Neurons by Appetitive Rather Than Aversive Stimuli," *Nature* 379 (1996): 449–451.

15. S. Lehéricy et al., "Diffusion Tensor Fiber Tracking Shows Distinct Corticostriatal Circuits in Humans," *Annals of Neurology* 55 (2004): 522–527.

16. E. R. Pickett et al., "Selective Speech Motor, Syntax, and Cognitive Deficits Associated with Bilateral Damage to the Putamen and the Head of the Caudate Nucleus: A Case Study," *Neuropsychology* 36 (1998): 173–188.

17. Flowers and Robertson devised the "Odd-Man-Out" test described below. It was used in the Everest study discussed in the first chapter. K. A. Flowers and C. Robertson, "The Effects of Parkinson's Disease on the Ability to Maintain a Mental Set," *Journal of Neurology, Neurosurgery, and Psychiatry* 48 (1985): 517–529.

18. See O. Monchi et al., "Cortical Activity in Parkinson Disease During Executive Processing Depends on Striatal Involvement," *Brain* 130 (2007): 233–244; O. Monchi et al., "Wisconsin Card Sorting Revisited: Distinct Neural Circuits Participating in Different Stages of the Task Identified by Event-Related Functional Magnetic Resonance Imaging," *Journal of Neuroscience* 21 (2001): 7739–7741; O. Monchi et al., "Functional Role of the Basal Ganglia in the Planning and Execution of Actions," *Annals of Neurology* 59 (2006): 257–264; F. Simard et al., "Frontostriatal Contributions to Lexical Set-Shifting," *Cerebral Cortex* 21 (2011): 1084–1093.

19. A. D. Patel, *Music, Language, and the Brain* (New York: Oxford University Press, 2008).

20. J. Talairach and P. Tournoux, *Coplanar Stereotaxic Atlas of the Human Brain* (Stuttgart: Thieme, 1988).

21. J. T. Devlin and R. A. Poldark, "In Praise of Tedious Anatomy," *Neuroimage* 37 (2007): 1033–1058, point out that the only way that you can be certain that the fMRI signal is coming from a particular region of prefrontal cortex (the front portion of the human cortex that's often identified as the "seat" of higher cognition) is to sacrifice the subject and cut up his or her brain—a no-brainer.

22. K. A. Amunts et al., "Broca's Area Revisited: Cytoarchitecture and Intersubject Variability," *Journal of Comparative Neurology* 412 (1999): 319–341.

Area 44, which usually is the cortical region associated with Broca, though it isn't, varied as much as ten times when only ten brains were sectioned after death.

23. Duncan and Owen found that this was the case. Subsequent fMRI studies keep finding similar effects in different parts of the brain. J. Duncan and A. M. Owen, "Common Regions of the Human Frontal Lobe Recruited by Diverse Cognitive Demands," *Trends in Neurosciences* 10 (2000): 475–483.

24. Oury Monchi's research group at the University of Alberta has published a series of studies that argue against neural circuits devoted solely to one narrow task, for example, syntax or sorting out colors or shapes. The Monchi group's fMRI studies are listed in this chapter's note 18.

25. S. A. Kotz et al., "Syntactic Language Processing: ERP Lesion Data on the Role of the Basal Ganglia," *Journal of the International Neuropsychological Society* 9 (2003): 1053–1060.

26. M. Just et al., "Brain Activation Modulated by Sentence Comprehension," *Science* 274 (1996): 114–116—one of the very first fMRI studies—showed that brains pull in more resources as task demand increases.

27. Wang and colleagues observed the same phenomena as Just, but in different parts of the brain and in a different task. As task difficulty increased in a mental arithmetic task, activity in ventrolateral prefrontal cortex (the lower frontal part of the cortex) increased. The same cortical area, as well as other cortical areas and subcortical structures, are active in tasks ranging from understanding the meaning of a sentence, to sorting out shapes, to changing the direction of a thought process, learning names, etc. J. Wang et al., "Perfusion Functional MRI Reveals Cerebral Blood Flow Pattern Under Psychological Stress," *Proceedings of the National Academy of Sciences* 102 (2005): 17804–17809.

28. Sanes and colleague's study looked at the neural bases of learning motor acts. J. Sanes et al., "Shared Neural Substrates Controlling Hand Movements in Human Motor Cortex," *Science* 268 (1999): 1775–1777. It led to projects that are constructing brain-controlled prostheses based on its findings and subsequent studies that have identified "matrisomes"—clusters of neurons in motor cortex that form as a person learns to execute a task automatically, whether it's typing, shifting gears, playing the violin, or walking. If you can tap into the appropriate matrisome and connect the

signal to a computer system that controls servomotors for mechanical hands, fingers, legs, etc., an injured person can "naturally"—by just doing it—type, pick up things, walk, etc.

29. F. Vargha-Khadem et al., "Neural Basis of an Inherited Speech and Language Disorder," *Proceedings of the National Academy of Sciences* 95 (1998): 12695–12700; C. S. Lai et al., "FOXP2 Expression During Brain Development Coincides with Adult Sites of Pathology in a Severe Speech and Language Disorder," *Brain* 126 (2001): 2455–2462; K. E. Watkins et al., "MRI Analysis of an Inherited Speech and Language Disorder: Structural Brain Abnormalities," *Brain* 125 (2002): 465–478.

30. S. E. Fisher et al., "Localization of a Gene Implicated in a Severe Speech and Language Disorder," *Nature Genetics* 18 (1998): 168–170.

31. C. S. Lai et al., "*FOXP2* Expression During Brain Development Coincides with Adult Sites of Pathology in a Severe Speech and Language Disorder," *Brain* 126 (2001): 2455–2462.

32. The Chimpanzee Sequencing and Analysis Consortium, "Initial Sequence of the Chimpanzee Genome and Comparison with the Human Genome," *Nature* 437 (2005): 69–87.

33. W. Enard et al., "A Humanized Version of *Foxp2* Affects Cortico–Basal Ganglia Circuits in Mice," *Cell* 137 (2009): 961–971; S. Reimers-Kipping et al., "Humanized *Foxp2* Specifically Affects Cortico-Basal Ganglia Circuits," *Neuroscience* 175 (2011): 75–84.

34. J. Krause et al., "The Derived *FOXP2* Variant of Modern Humans Was Shared with Neandertals," *Current Biology* 17 (2007): 908–1912.

35. M. Meyer et al., "A High-Coverage Genome Sequence from an Archaic Denisovian Individual," *Science* 338 (2012), doi:10.1126/science.1224344.

36. A large group of fossils dating back 400,000 years were found in a cave in Spain. Analysis of DNA recovered from their bones shows mixed Neanderthal and Denisovian characteristics. See Meyer et al., "A High-Coverage Genome Sequence."

37. G. Konopka et al., "Human-Specific Transcriptional Regulation of CNS Development Genes by *FOXP2*," *Nature* 462 (2009): 213–217; G. Konopka and T. F. Roberts, "Insights Into the Neural and Genetic Basis of Vocal Communication," *Cell* 6 (2016): 1269–1276; T. Maricic et al., "A Recent Evolutionary Change Affects a Regulatory Element in the Human *FOXP2* Gene," *Molecular Biology and Evolution* 25 (2013): 1257–1259.

38. D. Gokhman et al., "Reconstructing the DNA Methylation Maps of the Neandertal and the Denisovan," *Science* 344 (2014): 523–527; D. Gokhman et al., "Recent Regulatory Changes Shaped Human Facial and Vocal Anatomy," *bioRxiv* (February 8, 2017), doi:http:dx.doi.org/10.1101/106955. Gokhman and his colleagues obviously were not able to work with embryonic material, on which epigenetic factors have the greatest effect, from Neanderthals, Denisovans, and humans long dead, so their inferences on methylation factors are subject to some measure of uncertainty, but they are quite reasonable and similar to those derived for other independent studies that compare DNA recovered from the bones of long-dead hominins with living humans and apes with morphological differences.

5. WHAT WOULD DARWIN THINK ABOUT . . .

1. Charles Darwin, *On the Origin of Species* (1859; Cambridge, Mass.: Harvard University Press, 1964), 64.
2. Darwin, *On the Origin of Species*, 71.
3. Darwin, *On the Origin of Species*, 109.
4. Charles Darwin, *The Descent of Man, and Selection in Relation to Sex* (London: John Murray, 1871), 105.
5. Laurie Goldstein, in the *New York Times* (June 26, 2016).
6. N. Chomsky, *Language and Mind* (New York: HBJ, 1972), 97.
7. N. Chomsky, *The Science of Language* (Cambridge: Cambridge University Press, 2012), 105.
8. Chomsky, *The Science of Language*, 18.
9. R. C. Berwick et al., "Evolution, Brain, and the Nature of Language," *Trends in Cognitive Sciences* 17 (2013): 89–98; J. J. Bolhuis et al., "How Could Language Have Evolved?" *PLoS Biology* 12 (2014). Many animals, including dogs, possess some perhaps attenuated form of Merge. Juliane Kaminski noticed that the family dog, Rico, a border collie, brought the Kaminski children their toys. Border collies are generally thought to be an intelligent breed, but other "superdogs" who can learn the names of hundreds of words with hardly any tutoring, if at all, have since been studied by Kaminski and her colleagues. J. Kaminski, J. Call, and J. Fisher, "Word Learning in a Domestic Dog: Evidence for Fast Mapping," *Science* 240 (2004): 1676–1671.

Also overlooked by Chomsky and his colleagues is the potential of chimpanzees to learn and productively use about 150 words, if they don't have to talk. As discussed earlier, Alan and Beatrix Gardner raised chimpanzees from infancy with humans who communicated with them using American Sign Language. ASL, like Latin and many other languages, conveys grammatical distinctions such as the subject or object of a sentence using case markings, or inflections, which in ASL are signaled by hand placement and movement. R. A. Gardner and B. T. Gardner, "Teaching Sign Language to a Chimpanzee," *Science* 165 (1969): 664–672.

Washoe, the first chimpanzee who lived with the Gardners in their home in suburban Reno, Nevada, even invented a new "semantic referent," a new meaning for the word *dirty*. Washoe had learned, as human children do, that when she was told by means of ASL that she was *dirty*, she had soiled her diaper. At her birthday party, Washoe insulted Roger Fouts, then one of the graduate students involved in the project, by telling him he was *dirty*. Fouts had refused to give Washoe a second slice of birthday cake. Washoe also signed *dirty* at a rhesus monkey that bit her. Washoe also used ASL to refer to past events, communicate her desires, and comment on what she observed. William Stokoe, a distinguished professor at Gallaudet University, the center for the study of sign language in the United States, observed the Gardner chimpanzees in action and attested to their using ASL productively. Moreover, the "Loulis" chimpanzee experiment, described in detail by Roger and Deborah Fouts, showed that chimpanzees can transmit ASL from one generation to the next. A young chimpanzee named Loulis was placed with a group of older ASL-using chimpanzees from the Gardner project. Over the course of a year, the humans who cared for the chimpanzees talked to the chimpanzees instead of using ASL. Loulis learned to use ASL through interaction with his chimpanzee "family." R. S. Fouts, D. H. Fouts, and T. Van Cantfort, "The Infant Loulis Learns from Cross-Fostered Chimpanzees," in *Teaching Sign Language to Chimpanzees*, ed. R. A. Gardner et al. (Albany: SUNY Press, 1989), 280–292.

The chimpanzees in these projects never progressed beyond the level of competence achieved by three-year-old ASL-using children, but they were able to converse with one another and with humans, communicating their wants, referring to past events, and arguing. I witnessed the Gardner-raised chimpanzees arguing, using ASL, about who should get the red blanket instead of the blue one when blankets were distributed before bedtime. Noam Chomsky and his followers, of course, dispute these facts, most often citing Herbert Terrace's flawed project. The procedures and metrics that Terrace used, if applied to humans talking to one another, would have showed that we don't possess language. H. S. Terrace et al., "Can an Ape Create a Sentence?" *Science* 206 (1979): 821–901. A detailed discussion of Terrace's study and its deficiencies is in my book, *The Biology and Evolution of Language* (Cambridge, Mass.: Harvard University Press, 1984), 240–246. E. Hess, in *Nim Chimpsky: The Chimp Who Would Be Human* (New York: Bantam, 2008), focuses on Terrace's project, his affair with the graduate student working under him on the project, and how the chimpanzee Nim was disposed of when the project ended. Her book is the basis for a movie on Terrace and his Nim Chimpsky project.

10. Darwin, *On the Origin of Species*, 191.

11. D. E. Lieberman and R. C. McCarthy, "The Ontogeny of Cranial Base Angulation in Humans and Chimpanzees and Its Implications for Reconstructing Pharyngeal Dimensions," *Journal of Human Evolution* 36 (1999): 487–517.

12. P. Lieberman and R. C. McCarthy, "The Evolution of Speech and Language," in *Handbook of Paleoanthropology*, 2nd ed., ed. W. Henke and I. Tattersall (Berlin: Springer, 2015).

13. O. Marin, W. J. Smeets, and A. Gonzalez, "Evolution of the Basal Ganglia in Tetrapods: A New Perspective Based on Recent Studies in Amphibians," *Trends in Neurosciences* 21 (1998): 487–494.

14. C.-Y. Tseng, "An Acoustic Study of Tones in Mandarin," PhD diss., Brown University, 1981.

15. A. Fernald et al., "A Cross-Language Study of Prosodic Modifications in Mothers' and Fathers' Speech to Preverbal Infants," *Journal of Child Language* 16 (1989): 477–501.

16. D. Everett, *How Language Began* (New York: Norton, in press).

17. A. Lameira et al., "Vocal Fold Control Beyond the Species-Specific Repertoire in an Orangutan," *Scientific Reports* 6 (2016): 30315.

18. Marc Hauser, in his *Moral Minds: How Nature Designed a Universal Sense of Right and Wrong* (New York: Harper Collins/Ecco, 2006), claimed that the neural mechanisms that determine what's right and wrong are innate human attributes. Hauser ignores the fact that the genes of Iceland's population have remained stable since the end of the Viking age. The Vikings

were brutal savages, raiding much of Europe, and life in Iceland was violent. Iceland today has the lowest rate of violent crime in the world—same gene frequencies, different culture.

19. Pinker in a 2012 NPR broadcast. He repeated the claims in Dennis Dutton's *The Art Instinct: Beauty, Pleasure, and Human Evolution* (London: Bloomsbury, 2009). A visit to any major museum would have disabused them.

20. Darwin, *On the Origin of Species*, 61.

21. Charles Darwin, *The Autobiography of Charles Darwin, 1809–1882*, restored ed., ed. Nora Barlow (London: Collins, 1958), 87.

22. Darwin, *The Autobiography of Charles Darwin*, 87.

23. Darwin, *The Autobiography of Charles Darwin*, 93.

BIBLIOGRAPHY

Alexander, G. E., F. H. DeLong, and P. L. Strick. "Parallel Organization of Segregated Circuits Linking Basal Ganglia and Cortex." *Annual Review of Neuroscience* 9 (1986): 357–381.

Alexander, M. P, M. A. Naeser, and C. L. Palumbo. "Correlations of Subcortical CT Lesion Sites and Aphasia Profiles." *Brain* 110 (1987): 961–991.

Amunts, K., A. Schleicher, U. Burgel, et al. "Broca's Area Revisited: Cytoarchitecture and Intersubject Variability." *Journal of Comparative Neurology* 412 (1999): 319–341.

Anderson, M. L., and B. L. Finlay. "Allocating Structure to Function: The Strong Links Between Neuroplasticity and Natural Selection." *Frontiers in Human Neuroscience* (2014). doi:10.3389/fnhum.2013.00918.

Badre, D., and A. D. Wagner. "Computational and Neurobiological Mechanisms Underlying Cognitive Flexibility." *Procedures of the National Academy of Science of the United States* A 103 (2006): 7186–7190.

Bar-Gad, I., and H. Bergman. "Stepping Out of the Box: Information Processing in the Neural Networks of the Basal Ganglia." *Current Opinion in Neurobiology* 11 (2001): 689–695.

Berwick, R. C., A. D. Friederici, N. Chomsky, et al. "Evolution, Brain, and the Nature of Language." *Trends in Cognitive Sciences* 17 (2013): 89–98.

Behrmann, M., and D. C. Plaut. "Distributed Circuits, Not Circumscribed Centers, Mediate Visual Recognition." *Trends in Cognitive Science* (2013). http://dx.doi.org/10.1016/j.tics.2013.03.007.

Blöde, K. A. *Dr. F. J. Gall's Lehre über die Verrichtungen des Gehirns, nach dessen Dresden gehaltenen Vorlesungen in einer fasslichen Ordnung mit gewissenhafter Treue dargestellt*. 2nd ed. Dresden, 1806.

Bolhuis J. J., I. Tattersaall, N. Chomsky et al. "How Could Language Have Evolved?" *PLoS Biology* 12 (2014): e1001934. doi:10.1371/journal.pbio.1001934 PMID: 25157536.

Bordes, F. *Typologie du Paléolithique ancien et moyen.* Bordeaux: Delmas, 1961.

Bouhuys, A. *Breathing.* New York: Grune and Stratton, 1974.

Boyle, R. *A Defence of the Doctrine Touching the Spring and Weight of the Air . . .* London: Thomas Robinson, 1662.

Brainard, M. S., and A. J. Doupe. "Interruption of a Basal Ganglia–Forebrain Circuit Prevents Plasticity of Learned Vocalizations." *Nature* 404 (2000): 762–766.

Broca, P. "Nouvelle observation d'aphémie produite par une lésion de la moitié postérieure des deuxième et troisième circonvolutions frontales." *Bulletins Societe Anatomique De Paris,* 2nd ser., 6 (1861): 398–407.

Brodmann, K. "Beiträge zur histologischen Lokalisation der Grosshirnrinde. VII. Mitteilung: Die cytoarchitektonische Cortexgleiderung der Halbaffen (Lemuriden)." *Journal für Psychologie und Neurologie* 10 (1908): 287–334.

——. "Ergebnisse uber die vergleichende histologische Lokalisation der Grosshirnrinde mit besonderer Berucksichtigung des Stirnhirns." *Anatomischer Anzeiger,* suppl., 41 (1912): 157–216.

——. *VergleichendeLokalisationslehre der Groshirnrinde in iheren Prinzipien dargestellt auf Grund des Zellenbaues.* Leipzig: Barth, 1909.

Browne, J. *Charles Darwin: The Power of Place.* Princeton, N.J.: Princeton University Press, 2002.

——. *Charles Darwin, Voyaging.* Princeton, N.J.: Princeton University Press, 1995.

Bygren, L. O., P. Tinghog, J. Carstensen, et al. "Change in Paternal Grandmothers' Early Food Supply Influenced Cardiovascular Mortality of the Female Grandchildren." *BMC Genetics* 15 (2014). doi:10.1186/1471-2156 -15-12.

Carre R., B. Lindblom, and P. MacNeilage. "Acoustic Factors in the Evolution of the Human Vocal Tract." *Comptes Rendus de l'Académie des Sciences (IIb)* 320 (1995): 471–476.

Chambers, R. *Vestiges of the Natural History of Creation and Other Evolutionary Writings.* Chicago: University of Chicago Press, 1844/1994.

Cheyney, D. L., and R. M. Seyfarth. *How Monkeys See the World: Inside the Mind of Another Species.* Chicago: University of Chicago Press, 1990.

Chimpanzee Sequencing and Analysis Consortium. "Initial Sequence of the Chimpanzee Genome and Comparison with the Human Genome." *Nature* 437 (2005): 69–87.

Chomsky, C. *The Acquisition of Syntax in Children from Five to Ten*. Cambridge, Mass.: MIT Press, 1969.

Chomsky, N. *Language and Mind*. New York: HBJ, 1972.

——. *The Science of Language*. Cambridge: Cambridge University Press, 2012.

Combe, G. *The Constitution of Man and Its Relation to External Objects*. Edinburgh: Maclachlan, Stewart, & Co., etc., 1847.

Cooke, S. E., R. W. Komorowski, E. S. Kaplan, et al. "Visual Recognition Memory, Manifested as Long-Term Habituation Requires Synaptic Plasticity in V1." *Nature Neuroscience* 18 (2015): 262–271.

Coon, C. 1962. *The Origin of Races*. New York: Random House.

Corbey, R., A. Jagich, K. Vaesen, and M. Collard. "The Acheulean Handaxe: More Like a Bird's Song Than a Beatles Tune?" *Evolutionary Anthropology: Issues, News, and Reviews* 25 (2016): 6–19.

Crelin, E. S. *Anatomy of the Newborn: An Atlas*. Philadelphia: Lea and Febiger, 1969.

Cummings, J. L. "Frontal-Subcortical Circuits and Human Behavior." *Archives of Neurology* 50 (1993): 873–880.

Darwin, C. *The Autobiography of Charles Darwin, 1809–1882*. Restored ed. Ed. Nora Barlow. London: Collins, 1958.

——. *The Descent of Man, and Selection in Relation to Sex*. London: John Murray, 1871.

——. *Notebook E: Transmutation (1838–1839)*. http://darwin-online.org.uk /EditorialIntroductions/vanWyhe_notebooks.html

——. *On the Origin of Species*. 1859; Cambridge, Mass.: Harvard University Press, 1964.

——. 1845 (2nd rev. ed.); *The Voyage of the* Beagle. London: John Murray, 1905.

Darwin, E. *Zoonomia; or, The Laws of Organic Life*. 1794–1796. http://books .google.co.uk/books?id=A0gSAAAAYAAAJ.

Darwin, F., ed. *The Life and Letters of Charles Darwin, Including an Autobiographical Chapter*. London: John Murray, 1888.

De Boer, B. "Modeling Vocal Anatomy's Significant Effect on Speech." *Journal of Evolutionary Psychology* 8 (2010): 351–366.

De Waal, F. *Chimpanzee Politics: Power and Sex Among Apes*. Rev. ed. Baltimore, Md.: Johns Hopkins University Press, 2007.

Devlin, J. T., and R. A. Poldark. "In Praise of Tedious Anatomy." *Neuroimage* 37 (2007): 1033–1058.

Dominissini, D., C. He, and G. Rechavi. "RNA Epigenetics." *The Scientist* (2016). http://www.the-scientist.com/?articles.view/articleNo/44873/title/RNA -Epigenetics/.

Dronkers, N. F., O. Plaisant, M. T. Iba-Zizen, et al. "Paul Broca's Historic Cases: High-Resolution MR Imaging of the Brains of Leborgne and Lelong." *Brain* 130 (2007): 1432–1441.

Duncan, J., and A. M. Owen. "Common Regions of the Human Frontal Lobe Recruited by Diverse Cognitive Demands." *Trends in Neurosciences* 10 (2000): 475–483.

Dutton, D. *The Art Instinct: Beauty, Pleasure, and Human Evolution.* London: Bloomsbury, 2009.

Enard, W., S. Gehre, K. Hammerschmidt, et al. "A Humanized Version of *FOXP2* Affects Cortico–Basal Ganglia Circuits in Mice." *Cell* 137 (2009): 961–971.

Enard W., M. Przeworski, S. E. Fisher, et al. "Molecular Evolution of *FOXP2*, a Gene Involved in Speech and Language." *Nature* 41 (2002): 869–872.

Everett, D. *How Language Began.* New York: Norton, in press.

Federenko, E., M. K. Behr, and N. Kanwisher. "Functional Specificity for High-Level Linguistic Processing in the Human Brain." *Proceedings of the National Academy of Sciences* 108, no. 39 (2011).

Fernald, A., T. Taeschner, J. Dunn, et al. "A Cross-Language Study of Prosodic Modifications in Mothers' and Fathers' Speech to Preverbal Infants." *Journal of Child Language* 16 (1989): 477–501.

Finlay, B. L., and R. Uchiyama. "Developmental Mechanisms Channeling Cortical Evolution." *Trends in Neurosciences* (2014).

Fitch, W. T., B. de Boer, N. Mathur, et al. "Monkey Vocal Tracts Are Speech-Ready." *Science Advances* 2 (2016).

Flammang, B. E., A. Suvarnaraksha, J. Markiewicz, and D. Soares. "Tetrapod-like Pelvic Girdle in a Walking Cavefish." *Scientific Reports* (2016). doi:10 .1038/srep23711.

Flowers, K. A., and C. Robertson. "The Effects of Parkinson's Disease on the Ability to Maintain a Mental Set." *Journal of Neurology, Neurosurgery, and Psychiatry* 48 (1985): 517–529.

Fodor, J. A. 1983. *Modularity of Mind.* Cambridge, Mass.: MIT Press, 1983.

Fodor, J. A., and M. Piatelli-Palmarini. *What Darwin Got Wrong*. New York: FSG, 2011.

Forsdahl, A. "Are Poor Living Conditions in Childhood and Adolescence an Important Risk Factor for Arteriosclerotic Heart Disease?" *British Journal of Preventative and Social Medicine* 31 (1977): 91–95.

Fouts, R. S., and D. H. Fouts. "Chimpanzees' Use of Sign Language." In *The Great Ape Project*, ed. P. Cavalieri and P. Singer, 28–41. New York: St. Martin's Griffin, 1993.

Fouts, R. S., D. H. Fouts, and T. Van Cantfort. "The Infant Loulis Learns from Cross-Fostered Chimpanzees." In *Teaching Sign Language to Chimpanzees*, ed. R. A. Gardner et al., 280–292. Albany: SUNY Press, 1989.

Fouts, R. S., A. D. Hirsch, and D. H. Fouts. "Cultural Transmission of a Human Language in a Chimpanzee Mother-Infant Relationship." In *Child Nurturance*, vol. 3, ed. H. E. Fitzgerald et al. New York: Plenum, 1982.

Fumagalli M., L. Moltkee, N. Garup, et al. "Greenland Inuit Show Genetic Signatures of Diet and Climate Adaptation." *Science* 349 (2015): 1343–1347.

Gardner, R. A., and B. T. Gardner. "Teaching Sign Language to a Chimpanzee." *Science* 165 (1969): 664–672.

Gauthier, I., P. Skudlarski, J. C. Gore, and A. W. Anderson. "Expertise for Cars and Birds Recruits Brain Areas Involved in Face Recognition." *Nature Neuroscience* 3 (2000): 191–197.

Gauthier, I., and M. J. Tarr. "Unraveling Mechanisms for Expert Object Recognition: Bridging Brain Activity and Behavior." *Journal of Experimental Psychology: Human Perception and Performance* 28 (2002): 431–446.

Gavornik, J. P., and M. F. Bear. "Learned Spatiotemporal Sequence Recognition and Prediction in Primary Visual Cortex." *Nature Neuroscience* 17 (2014): 732–737.

Gokhman, D., E. Lavi, K. Prufer, et al. "Reconstructing the DNA Methylation Maps of the Neanderthal and the Denisovan." *Science* 344 (2014): 523–527.

Gokhman, D., L. Agranat-Tamir, G. Housman, et al. "Recent Regulatory Changes Shaped Human Facial and Vocal Anatomy." *bioRxiv* (February 8, 2017). doi:https://doi.org/10.1101/106955.

Goldstein, K. *Language and Language Disturbances: Aphasic Symptom Complexes and Their Significance for Medicine and Theory of Language*. New York: Grune, 1948.

Goodall, J. *The Chimpanzees of Gombe: Patterns of Behavior.* Cambridge, Mass.: Harvard University Press, 1986.

Gould, S. J., and R. C. Lewontin. "The Spandrels of San Marco and the Panglossian Paradigm: A Critique of the Adaptationist Programme." *Proceedings of the Royal Society of London Series B* 205, no. 1161 (1979): 581–598.

Graybiel, A. M. "The Basal Ganglia and Cognitive Pattern Generators." *Schizophrenia Bulletin* 23 (1997): 459–469.

——. "Building Action Repertoires: Memory and Learning Functions of the Basal Ganglia." *Current Opinion in Neurobiology* 5 (1995): 733–741.

Harmand, S. J., E. Lews, C. S. Feibel, et al. "3.3-Million-Year-Old Stone Tools from Lomekwi 3, West Turkana, Kenya." *Nature* 521 (2015): 310–315.

Hauser, M. D. *Moral Minds: How Nature Designed a Universal Sense of Right and Wrong.* New York: Harper Collins/Ecco, 2006.

Hebb, D. O. *The Organization of Behavior: A Neurophysiological Theory.* New York: Wiley, 1949.

Henshilwood, C., F. d'Errico, K. L. van Niekerk, et al. "A 100,000-Year-Old Ochre Processing Workshop at the Blombos Cave, South Africa." *Science* 334 (2011): 219–221.

Hess, E. *Nim Chimpsky: The Chimp Who Would Be Human.* New York: Bantam, 2008.

Hubel, D. H., and T. N. Wiesel. "Shape and Arrangement of Columns in Cat's Striate Cortex." *Journal of Physiology* 165 (1963): 559–568.

Jin, X., and R. M. Costa. "Start/Stop Signals Emerge in Nigrostriatal Circuits During Sequence Learning." *Nature* 466 (2010): 457–462.

Joshua, M., A. Adler, R. Mitelman, et al. 2008. "Midbrain Dopaminergic Neurons and Striatal Cholinergic Interneurons Encode the Difference Between Reward and Aversive Events at Different Epochs of Probabilistic Classical Conditioning Trials." *Journal of Neuroscience* 28 (2008): 11673–11684.

Just, M. A., P. A. Carpenter, T. A. Keller, et al. "Brain Activation Modulated by Sentence Comprehension." *Science* 274 (1996):114–116.

Kagan, J., S. Reznick, and N. Snidman. "Biological Bases of Childhood Shyness." *Science* 240 (1988): 167–171.

Kaminski, J., J. Call, and J. Fisher. "Word Learning in a Domestic Dog: Evidence for Fast Mapping." *Science* 240 (2004): 1676–1671.

Kanwisher, N., J. McDermott, and M. M. Chun. "The Fusiform Face Area: A Module in Human Extrastriate Cortex Specialized for Face Perception." *Journal of Neuroscience* 17 (1997): 4302–4311.

Kolata, A. L. *Ancient Inca*. New York: Cambridge University Press, 2013.

Konopka, G., J. M. Bomar, K., Winden, et al. "Human-Specific Transcriptional Regulation of CNS Development Genes by *FOXP2*." *Nature* 462 (2009): 213–217.

Konopka, G., and T. F. Roberts. "Insights Into the Neural and Genetic Basis of Vocal Communication." *Cell* 6 (2016): 1269–1276. doi:10.1016/j.cell.2016.02 .039.

Kosslyn, S. M., A. Pascual-Leone, O. Felician, et al. "The Role of Area 17 in Visual Imagery: Convergent Evidence from PET and rTMS." *Science* 284 (1999): 167–170.

Kotz, S. A., S. Frisch, D. Y. von Cramon, and A. D. Friederici. "Syntactic Language Processing: ERP Lesion Data on the Role of the Basal Ganglia." *Journal of the International Neuropsychological Society* 9 (2003): 1053–1060.

Krause, J., C. Lalueza-Fox, L. Orlando, et al. "The Derived *FOXP2* Variant of Modern Humans Was Shared with Neandertals." *Current Biology* 17 (2007): 908–1912.

Kuypers, H. "Corticobulbar Connections to the Pons and Lower Brainstem in Man." *Brain* 81 (1958): 364–388.

Lai, C. S., D. Gerrelli, A. P. Monaco, et al. "*FOXP2* Expression During Brain Development Coincides with Adult Sites of Pathology in a Severe Speech and Language Disorder." *Brain* 126 (2001): 2455–2462.

Lamarck, J.-B. *Philosophie zoologique*. Paris: Museum d'Histoire Naturelle, 1809.

Lameira, A., M. E. Hardus, A. Mielke, et al. "Vocal Fold Control Beyond the Species-Specific Repertoire in an Orangutan." *Scientific Reports* 6 (2016): 30315.

Lartet, E. "De quelques cas de progression organique vérifiables dans la succession des temps, géologiques sur des mammifères de même famille et de même genre." *Comptes Rendus de l'Académie des Sciences Paris* 66 (1868): 1119–1122.

Leakey, M. D. *Olduvai Gorge: Excavations in Beds I and II, 1960–1963*. Cambridge: Cambridge University Press, 1971.

Leal, M., and B. J. Powell. "Behavioural Flexibility and Problem Solving in a Tropical Lizard." *Biology Letters* 8 (2012): 28–30.

Lehéricy, S., M. Ducros, P. F. Van de Moortele, et al. "Diffusion Tensor Fiber Tracking Shows Distinct Corticostriatal Circuits in Humans." *Annals of Neurology* 55 (2004): 522–527.

Liberman, A. M., F. S. Cooper, D. P. Shankweiler, et al. "Perception of the Speech Code." *Psychological Review* 74 (1967): 431–461.

Lichtheim, L. "On Aphasia." *Brain* 7 (1885): 433–484.

Lieberman, D. E. *The Evolution of the Human Head*. Cambridge, Mass.: Harvard University Press, 2011.

——. *The Story of the Human Body: Evolution, Health, and Disease*. New York: Pantheon, 2013.

Lieberman, D. E., and R. C. McCarthy. "The Ontogeny of Cranial Base Angulation in Humans and Chimpanzees and Its Implications for Reconstructing Pharyngeal Dimensions." *Journal of Human Evolution* 36 (1999): 487–517.

Lieberman, P. *The Biology and Evolution of Language*. Cambridge, Mass.: Harvard University Press, 1984.

——. *Human Language and Our Reptilian Brain: The Subcortical Bases of Speech, Syntax, and Thought*. Cambridge, Mass.: Harvard University Press, 2000.

——. *Intonation, perception, and language*. Cambridge, Mass.: MIT Press, 1967.

——. "On the Nature and Evolution of the Neural Bases of Human Language." *Yearbook of Physical Anthropology* 45 (2002): 36–62.

——. *Toward an Evolutionary Biology of Language*. Cambridge, Mass.: Harvard University Press, 2006.

——. *The Unpredictable Species: What Makes Humans Unique*. Princeton, N.J.: Princeton University Press, 2013.

——. "Vocal Tract Anatomy and the Neural Bases of Talking." *Journal of Phonetics* 40 (2012): 608–622.

Lieberman, P., and E. S. Crelin. "On the Speech of Neanderthal Man." *Linguistic Inquiry* 2 (1971): 203–222.

——. "Comment on 'Monkey Vocal Tracts Are Speech Ready'; Monkey Business: Did the Evolution of Speech Involve Anatomy?" *Science Advances* 3 (2017): e1700442.

Lieberman, P., E. S. Crelin, and D. H. Klatt. "Phonetic Ability and Related Anatomy of the Newborn, Adult Human, Neanderthal Man, and the Chimpanzee." *American Anthropologist* 74 (1972): 287–307.

Lieberman, P., E. T. Kako, J. Friedman, et al. "Speech Production, Syntax Comprehension, and Cognitive Deficits in Parkinson's Disease." *Brain Language* 43 (1992): 169–189.

Lieberman, P., B. G. Kanki, A. Protopapas, et al. "Cognitive Defects at Altitude." *Nature* 372 (1994): 325.

Lieberman, P., D. H. Klatt, and W. H. Wilson. "Vocal Tract Limitations on the Vowel Repertoires of Rhesus Monkey and Other Nonhuman Primates." *Science* 164 (1969): 1185–1187.

Lieberman, P., A. Morey, J. Hochstadt, et al. "Mount Everest: A Space-Analog for Speech Monitoring of Cognitive Deficits and Stress." *Aviation, Space, and Environmental Medicine* 76 (2005): 198–207.

Lieberman, P., and R. C. McCarthy. "The Evolution of Speech and Language." In *Handbook of Paleoanthropology*, 2nd ed., ed. W. Henke and I. Tattersall. Berlin: Springer, 2015.

——. "Tracking the Evolution of Language and Speech." *Expeditions* 49 (2007): 15–20.

Lisker, L., and A. S. Abramson. "A Cross-Language Study of Voicing in Initial Stops: Acoustical Measurements." *Word* 20 (1964): 384–442.

Lordkipanidze, D., M. S. Ponce de Leòn, A. Margvelashvili, et al. "A Complete Skull from Dmanisi, Georgia, and the Evolutionary Biology of Early *Homo*." *Science* 342 (2013): 326–331. doi:10.1126/science.1238484.

Lyko F., S. Foret, R. Kucharski, et al. "The Honey Bee Epigenomes: Differential Methylation of Brain DNA in Queens and Workers." *PLoS Biology* (2010).

Lyko, F., B. H. Ramsahoye, and R. Jaenisch. "DNA Methylation in *Drosophila melanogaster*." *Nature* 408 (2000): 538–540.

MacLean, P. D., and J. D. Newman. "Role of Midline Frontolimbic Cortex in the Production of the Isolation Call of Squirrel Monkeys." *Brain Research* 450 (1988): 111–123.

Maricic T., V. Günther, O. Georgiev, et al. "A Recent Evolutionary Change Affects a Regulatory Element in the Human *FOXP2* Gene." *Molecular Biology and Evolution* 25 (2013): 1257–1259.

Marie, P. *Traveaux et mémoires*. Paris: Masson, 1926.

Marin, O., W. J. Smeets, and A. Gonzalez. "Evolution of the Basal Ganglia in Tetrapods: A New Perspective Based on Recent Studies in Amphibians." *Trends in Neurosciences* 21 (1998): 487–494.

Marsden, C. D., and J. A. Obeso. "The Functions of the Basal Ganglia and the Paradox of Sterotaxic Surgery in Parkinson's Disease." *Brain* 117 (1994): 877–897.

Mayr, E. Introduction to *On the Origin of Species*, by Charles Darwin. Facsimile ed. Cambridge, Mass.: Harvard University Press, 1964.

McBrearty, S., and A. Brooks. "The Revolution That Wasn't: A New Interpretation of the Origin of Modern Human Behavior." *Journal of Human Evolution* 39 (2000): 453–563.

Meckel, J. F. *Beytrage zur Gechiichte des menschlichen Fotus, Beytrage vur vergleichenden Anatomie.* Leipzig: Reclam, 1808.

Mercader, J., H. Barton, J. Harris, et al. "4,300-Year-Old Chimpanzee Site and the Origins of Percussive Stone Technology." *Proceedings of the National Academy of Sciences* 104 (2007): 3043–3048.

Meyer, M., et al. "A High-Coverage Genome Sequence from an Archaic Denisovian Individual." *Science* 338 (2012). doi:10.1126/science.1224344.

Meyer, M., J.-L. Arsuaga, C. de Filippo, et al. "Nuclear DNA Sequences from the Middle Pleistocene Sima de los Huesos Hominins." *Nature* 531 (2016): 504–507. doi:10.1038/nature17405.

Miller, E. K., and M. A. Wilson. "All My Circuits: Using Multiple Electrodes to Understand Functioning Neural Networks." *Neuron* 60 (2008): 483–488.

Mirenowicz, J., and W. Schultz. "Preferential Activation of Midbrain Dopamine Neurons by Appetitive Rather Than Aversive Stimuli." *Nature* 379 (1996): 449–451.

Monchi, O., M. Petrides, B., Mejia-Constain, et al. "Cortical Activity in Parkinson Disease During Executive Processing Depends on Striatal Involvement." *Brain* 130 (2007): 233–244.

Monchi, O., M. Petrides, V. Petre, et al. "Wisconsin Card Sorting Revisited: Distinct Neural Circuits Participating in Different Stages of the Task Identified by Event-Related Functional Magnetic Resonance Imaging. *Journal of Neuroscience* 21 (2001): 7739–7741.

Monchi, O., M. Petrides, A. P. Strafella, et al. "Functional Role of the Basal Ganglia in the Planning and Execution of Actions." *Annals of Neurology* 59 (2006): 257–264.

Mouse Genome Sequencing Consortium. "Initial Sequencing and Comparative Analysis of the Mouse Genome." *Nature* 420 (2002): 520–562.

Nearey, T. *Phonetic Features for Vowels.* Bloomington: Indiana University Linguistics Club, 1978.

Negus, V. E. *The Comparative Anatomy and Physiology of the Larynx.* New York: Hafner, 1949.

Paley, W. *Natural Theology; or, Evidences of the Existence and Attributes of the Deity.* London: R. Faulder, 1802.

Pembrey, M. E., L. O. Bygren, G. Kaati, et al. "Sex-Specific, Male-Line Transgenerational Responses in Humans." *European Journal of Human Genetics* 14 (2006): 159–166.

Pembrey, M. E., R. Saffeertm, L. O. Bygren, and Network in Epigenetic Epidemiology. "Human Transgenerational Responses to Early-Life Experience: Potential Impact on Development, Health, and Biomedical Research." *Medical Genetics* 9 (2014): 563–572.

Peterson, G. E., and H. L. Barney. "Control Methods Used in a Study of the Vowels." *Journal of the Acoustical Society of America* 24 (1952): 175–184.

Petrides, M. "Lateral Prefrontal Cortex: Architectonic and Functional Organization." *Philosophical Transactions of the Royal Society B* 360 (2005): 781–795.

Pickett, E. R., E. Kuniholm, A. Protopapas, et al. "Selective Speech Motor, Syntax, and Cognitive Deficits Associated with Bilateral Damage to the Putamen and the Head of the Caudate Nucleus: A Case Study." *Neuropsychology* 36 (1998): 173–188.

Pinker, S. *How the Mind Works*. New York: Norton, 1998.

——. *The Language Instinct: How the Mind Creates Language*. New York: William Morrow, 1994.

Postle, B. R. "Working Memory as an Emergent Property of the Mind and Brain." *Neuroscience* 139 (2006): 23–38.

Rasmussen, S., M. E. Allentoft, K. Nielson, et al. "Early Divergent Strains of *Yersinia pestis* in Eurasia 5,000 Years Ago." *Cell* 163 (2015): 571–582.

Reilly, S. K., J. Yin, A. E. Ayoub, et al. "Evolutionary Changes in Promotor and Enhancer Activity During Human Corticogenesis." *Science* 347 (2015): 1155–1159.

Reynolds, V., et al. "Mineral Acquisition from Clay by Budongo Forest Chimpanzees." *PLoS One* (July 28, 2015). http://dx.doi.org/10.1371/journal.pone .0134075.

Reimers-Kipping, S., S. Hevers, S. Paabo, and W. Enard. "Humanized Foxp2 Specifically Affects Cortico-Basal Ganglia Circuits." *Neuroscience* 175 (2011): 75–84.

Sam, F., S. F. Cooke, R. W. Komorowski, et al. "Visual Recognition Memory, Manifested as Long-Term Habituation, Requires Synaptic Plasticity in V1." *Nature Neuroscience* 18 (2015): 262–273.

Sanes, J. N., J. P. Donoghue, V. Thangaraj, et al. "Shared Neural Substrates Controlling Hand Movements in Human Motor Cortex." *Science* 268 (1999): 1775–1777.

Savage-Rumbaugh, S., D. Rumbaugh, and K. McDonald. "Language Learning in Two Species of Apes." *Neuroscience and Biobehavioral Reviews* 9 (1985): 653–665.

Shea, J. J. "*Homo sapiens* Is as *Homo sapiens* Was: Behavioral Variability Versus 'Behavioral Modernity' in Paleolithic Archaeology." *Current Anthropology* 52 (2011): 1–34.

———. *Stone Tools in Human Evolution: Behavioral Differences Among Technological Primates.* New York: Cambridge University Press, 2017.

———. *Stone Tools in the Paleolithic and Neolithic of the Near East: A Guide.* New York: Cambridge University Press, 2013.

Shankweiler, D., L. C. Palumbo, W. Ni, et al. "Unexpected Recovery of Language Function After Massive Left-Hemisphere Infarct: Coordinated Psycholinguistic and Neuroimaging Studies." *Brain and Language* 91 (2004): 181–182.

Simard, F., Y. Joanette, M. Petrides, et al. "Frontostriatal Contributions to Lexical Set-Shifting." *Cerebral Cortex* 21 (2011): 1084–1093.

Simonson, T. S., Y. Yang, C. D. Huff, et al. "Genetic Evidence for High-Altitude Adaptation in Tibet." *Science* 329 (2010): 71–74.

Smith, B. L. "Temporal Aspects of English Speech Production: A Developmental Perspective." *Journal of Phonetics* 6 (1978): 37–68.

Spelke, E. S., K. Breinlager, J. Macomber, et al. "Origins of Knowledge." *Psychological Review* 99 (1992): 605.

Spurzheim, J. K. *The Physiognomical System of Drs. Gall and Spurzheim.* London: Baldwin, Cradock, and Joy, 1815.

Stahl, A., and E. Feigenson. "Observing the Unexpected Enhances Infants' Learning and Exploration." *Science* 348 (2015): 91–97.

Stevens, K. N. "Quantal Nature of Speech." In *Human Communication: A Unified View,* ed. E. E. David Jr. and P. B. Denes. New York: McGraw Hill, 1972.

Stukeley, W. *Memoirs of Sir Isaac Newton's Life.* 1752. http://www.royalsociety.org/turning-the-page.

Stuss, D. T., and D. F. Benson. *The Frontal Lobes.* New York: Raven, 1986.

Talairach, J., and P. Tournoux. *Coplanar Stereotaxic Atlas of the Human Brain.* Stuttgart: Thieme, 1988.

Terrace, H. S., L. A. Petitto, R. J. Sanders, and T. G. Bever. "Can an Ape Create a Sentence?" *Science* 206 (1979): 821–901.

Thelen, E. "Learning to Walk: Ecological Demands and Phylogenetic Constraints." In *Advances in Infancy Research*, ed. L. Lipsitt, 3:213–250. Norwood, N.J.: Ablex, 1984.

Thompson, D'arcy W. *On Growth and Form*. 1917; New York: Cambridge University Press, 1945.

Tishkoff, S. A., F. Reed, et al. "Convergent Adaptation of Human Lactase Persistence in Africa and Europe." *Nature Genetics* 39 (2007): 31–40.

Truby, H. L, J. F. Bosma, and J. Lind. *Newborn Infant Cry*. Uppsala: Almquist and Wiksell, 1965.

Tseng, C.-Y. "An Acoustic Study of Tones in Mandarin." PhD diss., Brown University, 1981.

Tyson, E. *Orang-outang, sive, Homo sylvestris; or, The anatomy of a pygmie compared with that of a monkey, an ape, and a man*. London: Printed for Thomas Bennet and Daniel Brown, 1699.

von Baer, K. E. *Über Entwickelungsgeschichte der Thiere*. Koningsberg: Borntrager, 1828.

Vargha-Khadem, F., K. Watkins, K. Alcock, et al. "Praxic and Nonverbal Cognitive Deficits in a Large Family with a Genetically Transmitted Speech and Language Disorder." *Proceedings of the National Academy of Sciences* 92 (1995): 930–933.

Vargha-Khadem, F., K. E. Watkins, C. J. Price, et al. "Neural Basis of an Inherited Speech and Language Disorder." *Proceedings of the National Academy of Sciences* 95 (1998): 12695–12700.

Wallaae, A. R. "On the Tendency of Varieties to Depart Indefinitely from the Original Type." *Proceedings of the Linnean Society of London* 3 (1858): 53–62.

Wang J., H. Rao, G. S. Wetmore, et al. "Perfusion Functional MRI Reveals Cerebral Blood Flow Pattern Under Psychological Stress." *Proceedings of the National Academy of Sciences* 102 (2005): 17804–17809.

Watkins, K. E., F. Vargha-Khadem, J. Ashburner, et al. "MRI Analysis of an Inherited Speech and Language Disorder: Structural Brain Abnormalities." *Brain* 125 (2002): 465–478.

Weiner, J. *The Beak of the Finch: A Story of Evolution in Our Time*. New York: Knopf, 1994.

Wernicke, C. "The Aphasic Symptom Complex: A Psychological Study on a Neurological Basis." 1874. In *Proceedings of the Boston Colloquium for the*

Philosophy of Science, vol. 4, ed. R. S. Cohen and M. W. Wartofsky. Dordrecht: Reidel, 1967.

Wolpoff, M., J. D. Hawks, and R. Caspari. "Multiregional, Not Multiple Origins." *American Journal of Physical Anthropology* 112 (2000): 129–136.

Zilhãoa, J., D. E. Angeluccib, E. Badal-Garcíac, et al. "Symbolic Use of Marine Shells and Mineral Pigments by Iberian Neanderthals." *Proceedings of the National Academy of Sciences* 107 (2010): 1023–1028.

INDEX